FIRST EDITION

FUNDAMENTALS OF CALCULUS FOR TEACHERS

DUSTIN JONES
Sam Houston State University

Bassim Hamadeh, CEO and Publisher
Seidy Cruz, Specialist Acquisitions Editor
Gem Rabanera, Project Editor
Chelsey Schmid, Production Supervisor
Jess Estrella, Senior Graphic Designer
Trey Soto, Licensing Coordinator
Natalie Piccotti, Director of Marketing
Kassie Graves, Vice President of Editorial
Jamie Giganti, Director of Academic Publishing

Copyright © 2020 by Dustin Jones. All rights reserved. No part of this publication may be reprinted, reproduced, transmitted, or utilized in any form or by any electronic, mechanical, or other means, now known or hereafter invented, including photocopying, microfilming, and recording, or in any information retrieval system without the written permission of Cognella, Inc. For inquiries regarding permissions, translations, foreign rights, audio rights, and any other forms of reproduction, please contact the Cognella Licensing Department at rights@cognella.com.

Trademark Notice: Product or corporate names may be trademarks or registered trademarks, and are used only for identification and explanation without intent to infringe.

Cover image copyright © 2014 Depositphotos/agsandrew.

Printed in the United States of America.

CONTENTS

Chapter 0   Introduction — 1

Think — 1
Remember — 1
Connect — 1
0.1 To the Instructor — 2
   0.1.1 The Need for This Course — 2
   0.1.2 Using This Book — 2
   0.1.3 Prerequisite Knowledge and Experiences — 3
0.2 To the Student — 3
   0.2.1 An Interesting Equation — 4
   0.2.2 Change — 4
   0.2.3 Area — 5
0.3 Technology Tools — 5
   Problem Set 0.3 — 6
References — 6

Chapter 1   Limits: Concepts and Applications — 7

Think — 7
Remember — 7
Connect — 8
1.1 Sequences and Series — 8
   Problem Set 1.1 — 9
1.2 Infinite Geometric Series — 10
   Problem Set 1.2 — 13
1.3 Repeating Decimals — 14
   Problem Set 1.3 — 16
1.4 Perimeter of a Circle — 16
   1.4.1 Limit Notation — 16
   1.4.2 Measuring Around a Circle — 17
   Problem Set 1.4 — 18
1.5 Location of a Hole in a Graph — 20
   Problem Set 1.5 — 21
References — 22

Chapter 2 Derivatives: Concepts — 23

Think — 23
Remember — 23
Connect — 24
2.1 How Do We Measure Speed? — 24
  Problem Set 2.1 — 30
2.2 The Derivative at a Point — 32
  Problem Set 2.2 — 37
2.3 The Derivative Function — 39
  Problem Set 2.3 — 44
2.4 Interpretations of the Derivative — 46
  Problem Set 2.4 — 49
2.5 The Second Derivative — 50
  Problem Set 2.5 — 53
References — 54

Chapter 3 Integrals: Concepts — 55

Think — 55
Remember — 55
Connect — 56
3.1 How Do We Measure Distance Traveled? — 56
  Problem Set 3.1 — 61
3.2 The Definite Integral — 62
  Problem Set 3.2 — 67
3.3 Area Between Curves — 70
  Problem Set 3.3 — 75
3.4 Interpretations of the Definite Integral — 77
  Problem Set 3.4 — 81
3.5 Theorems About Definite Integrals — 83
  3.5.1 Investigations and Explorations — 83
  3.5.2 Illustrations and Explanations — 85
  Problem Set 3.5 — 90
References — 91

Chapter 4 Derivatives and Integrals: Applications with Formulas — 93

Think — 93
Remember — 93
Connect — 94
4.1 Derivatives of Powers and Polynomials — 94
  Problem Set 4.1 — 100
4.2 Derivatives of Exponential Functions — 101
  Problem Set 4.2 — 105
4.3 Derivatives of Composite Functions — 106
  Problem Set 4.3 — 110

4.4 Antiderivatives	111
Problem Set 4.4	117
4.5 The Fundamental Theorem of Calculus	120
Problem Set 4.5	126
References	127

Chapter 5 Limits, Derivatives, and Integrals: Contexts Within and Beyond Middle School — 129

Think	129
Remember	129
Connect	129
5.1 Area of a Circle	130
Problem Set 5.1	132
5.2 Extrema	133
Problem Set 5.2	137
5.3 Optimization	138
Problem Set 5.3	141
5.4 Length of a Curve	142
Problem Set 5.4	145
5.5 Solids of Revolution: Volume and Surface Area	146
5.5.1 Solids of Revolution	146
5.5.2 Volume of a Solid of Revolution	147
5.5.3 Surface Area of a Solid of Revolution	151
Problem Set 5.5	153
References	154

Introduction

Think

What are you looking for in a textbook? The answer to this question depends on who you are and what you plan to do with it. In this introductory chapter, I will describe the purpose of this book to two audiences: those who will use the book as an *instructor* and those who will use the book as a *student*.

Remember

At the beginning of each chapter, I'll describe some prerequisite knowledge and brief mathematics exercises that will come in to play during the chapter. To prepare for this chapter, you may want to think of some different ways that you have used technology in the past to learn and do mathematics.

Connect

The major purpose of this textbook is to connect concepts from calculus to the mathematics content that is taught in the middle grades (e.g., grades 4–8 in the United States). In each chapter, I'll mention these topics. In this introduction, you'll get a glimpse of three topics: repeating decimals, rates of change, and area.

0.1 To the Instructor

OBJECTIVES FOR SECTION 0.1: Upon completing this section, you will understand the following:

- The rationale for creating this book
- What this book is (and what it is not)
- The prerequisite knowledge needed for this course

0.1.1 The Need for This Course

Middle grades mathematics teachers (MGMTs) (those who teach mathematics to students in grades 4 through 8) are uniquely positioned between the elementary and secondary levels. They must understand the mathematical content from both lower and higher grades, build on the mathematical knowledge that students bring with them, and prepare them for study of more advanced and abstract topics. This has implications for the preparation of prospective MGMTs. Both the *Mathematical Education of Teachers* (MET) (Conference Board of the Mathematical Sciences [CBMS], 2001) and the *Mathematical Education of Teachers II* (MET2) (CBMS, 2012) provide recommendations for mathematics courses focusing on two types of content: (a) those that relate specifically to the content that is taught in the middle grades (number and operations, geometry and measurement, algebra and number theory, and statistics and probability) and (b) those that focus on the content that students may encounter in later grades or levels. This second type of course may serve to extend the mathematical understanding of a prospective MGMT and develop horizon content knowledge (Ball & Bass, 2009; Ball, Thames, & Phelps, 2008).

A specific course of the second type, mentioned in both the MET and MET2, is a one-semester course that focuses on concepts and applications of calculus. According to the CBMS (2001), "The usual calculus designed for engineers and mathematics majors would probably not have this focus because of its emphasis on connections with physics and engineering" (p. 118). Furthermore, "such a course could include a careful study of the concepts underlying the standard topics of calculus" (p. 48). Such recommendations have been taken seriously, as the NCTM CAEP Standards 2012 (National Council of Teachers of Mathematics, 2012) specifically state that all MGMTs should know a number of topics related to calculus, such as limits, rates of change, and differentiation and integration. In order to receive accreditation through this organization, teacher preparation programs must demonstrate how their programs for MGMTs address these standards.

0.1.2 Using This Book

My university offers a 27-credit-hour minor in mathematics for prospective MGMTs that includes a one-semester course focused on calculus concepts. This course is offered only for MGMTs and has been offered for over two decades. A major goal of the course is to relate mathematical content taught in the middle grades to its underpinnings from calculus. Therefore, major topics include sequences, series, limits, rates of change, and measures such as areas, lengths, surface areas, and volumes. Recently, I redesigned the course to include a significant technological component using spreadsheets, graphing calculators (both handheld and online, e.g., Desmos), and GeoGebra. This book is a product of that course redesign, and it has been used at multiple universities across the United States. Over the past two semesters of implementation, we have witnessed prospective MGMTs that are becoming comfortable with using various technological tools for learning mathematics, and we believe that such tools allow for a focus on concepts (such as the derivative as a rate of change) instead of procedural techniques (such as the product rule).

This book is intended to be used in conjunction with technology. (See section 0.3 for specific information.) At my university, class meetings are held in a computer lab so that students can access the Internet (i.e., videos and mathematical technology) to support their learning. At other universities, students bring laptop computers to class. The book is designed for students to work through a set of exercises, punctuated by guidance and input from the instructor. Students may work independently or in small groups, depending on the exercise and the instructor's preference. Throughout many of the exercises, technology is used to support student discovery as they search for patterns, make and test conjectures, form generalizations, and draw conclusions.

Each section concludes with a problem set. These problem sets refer to the content presented in the section, although some problems will challenge students to think deeply about the concepts presented. At times, problems from one section will lead into the content of the following section. These problem sets are brief, and I recommend assigning *every* problem.

While the primary intended audience of this book is prospective mathematics teachers in the middle grades, this book is also useful for anyone looking to improve their conceptual knowledge of calculus concepts. It has been used by high school calculus teachers as a form of professional development and could also be used to introduce high school students to concepts of calculus. Finally, this book could also be used for independent study by a motivated reader.

While this book does address the fundamental ideas of calculus, it is not intended as a substitute for the full calculus sequence traditionally taken by those majoring in mathematics. This book does not address the full range of functions, methods, and techniques that calculus affords. For this reason, those seeking to use or teach calculus as a career would benefit from further study of formal calculus.

0.1.3 Prerequisite Knowledge and Experiences

Prior to taking this course, students need to have experiences with common families of functions (linear, quadratic, polynomial, exponential, logarithmic) and transformations of functions represented in graphs, tables, and formulas. It will also be useful for students to be familiar with inverses of functions. Some symbolic algebraic manipulation is required, but not beyond the level of high school mathematics.

0.2 To the Student

OBJECTIVES FOR SECTION 0.2: Upon completing this section, you will be introduced to the following:

- Ways calculus relates to middle-grades mathematics
- The organization of this book

In this course, we will discuss the ideas of calculus and how they are related to mathematical topics that are taught in grades 4–8 in the United States.

Calculus is a broad subject, and there are many techniques and applications in many disciplines. We will not attempt to spend too much time on specific techniques. Instead, we will focus on broad concepts: limits, derivatives, and integrals.

All three of these topics are related. In fact, limits are used to determine derivatives and integrals! The crux of the matter for limits, derivatives, and integrals, and all of calculus, is that we are interested in determining the value of an unknown quantity. We do so by making approximations and then refining the approximations in various ways.

0.2.1 An Interesting Equation

Think about how we can make sense of this: $1 = 0.99999999\ldots$

No, I am not saying the number on the right side would round up to 1. It is, in fact, **exactly** equal to 1. Or, to put it another way, $0.99999999\ldots$ is another way of writing the number 1.

Not convinced? Think about it this way:

$$\begin{aligned} \frac{1}{3} &= 0.33333\ldots \\ +\frac{2}{3} &= 0.66666\ldots \\ \hline \frac{3}{3} &= 1 = 0.99999\ldots \end{aligned}$$

What does that have to do with calculus? Well, if we were making a line, it may seem that if you keep adding positive lengths (even if they get smaller as you go) you'll get an infinitely long line. Except in this case, we do not. As it turns out, calculus ideas help us make sense of this. In fact, in the first part of this class, we'll investigate the mathematics behind this in some detail. The ideas we'll look at are very powerful.

0.2.2 Change

Here's a problem similar to some found in an eighth-grade mathematics textbook: *Lydia's car will travel 183 miles on 6.1 gallons of fuel. How many gallons will be needed to travel 510 miles?*

> **Box 0.1: Possible Pitfalls**
>
> In problems like these, it is usually a good idea to consider whether the rate is constant.

This is a problem that deals with a rate—in this case, the rate of gasoline consumption. The problem is simplified somewhat in that the rate of gasoline consumption appears to be constant (30 miles per gallon). In practice, gasoline consumption varies according to the car's speed and whether the car is accelerating or decelerating.

If a rate is constant, the mathematics of linear functions we learn in eighth grade might be sufficient, but that is not what happens in the real world—rates vary, and the functions that describe them are not linear. Calculus gives us a way to deal with these more complicated situations. We will see how this happens in chapter 2.

0.2.3 Area

One topic that is covered in middle school is area, in particular, the area of a circle. The formula is, of course, $A = \pi r^2$, where r is the radius of the circle.

Did you ever wonder why this formula works? Area is always measured in square units and circles are far from square! Here's a video (Mathematicsonline, 2011) that helps explain the formula. Take a look!

https://www.youtube.com/watch?v=YokKp3pwVFc

The main idea in the video is that the circle is cut up into more and more wedges—so much so that the wedges start to look like triangles and the new figure looks more and more like a rectangle. Compared to other shapes, it is easy to find the area of a rectangle.

The idea of finding the area of a non-standard figure by cutting it up into smaller and smaller pieces becomes more formalized in calculus. These ideas will be covered in chapters 3 and 5.

0.3 Technology Tools

OBJECTIVES FOR SECTION 0.3: Upon completing this section, you will be able to do the following:

- Identify the technology tools used in this course
- Locate these tools for use inside and outside of class

In this course, we will use technology to help illuminate and explore the concepts of calculus. Specifically, we will use four different types at various times and in different ways.

1. A graphing calculator, such as TI-84+ or TI-Nspire
2. A spreadsheet, such as Microsoft Excel or Google Sheets
3. Desmos, an online graphing calculator available for free from www.desmos.com
4. GeoGebra, a mathematics software package available online or for free download from www.geogebra.org

> **Box 0.2: Teaching Tips**
>
> Students tend to learn new material best when it is presented in concrete examples. Hands-on materials and visuals are very useful. Technology can help students explore concepts dynamically.

As you progress through this textbook, you will be directed to use the Internet to watch short videos. It is important that you actually watch the videos, as they help provide a context for the activities, exercises, or problems that follow. You will also be asked to use spreadsheets, GeoGebra, and Desmos to create and explore mathematical ideas. Sometimes, you will be given instructions on how to use the technology to create your tools. At other times, you will use files that have already been prepared. Specifically, a collection of GeoGebra applets designed for this course is available at https://ggbm.at/BN8D2duN

Problem Set 0.3

1. Locate a computer that you can use outside of class. Be sure that you can run the following programs:
    a. A spreadsheet, such as http://sheets.google.com
    b. Desmos, at http://www.desmos.com
    c. GeoGebra, at http://www.geogebra.org

References

Ball, D.L., & Bass, H. (2009). *With an eye on the mathematical horizon: Knowing mathematics for teaching to learners' mathematical futures.* Paper presented at the 43rd Jahrestagung für Didaktik der Mathematik, Oldenburg, Germany.

Ball, D.L., Thames, M.H., & Phelps, G. (2008). Content knowledge for teaching: What makes it special? *Journal of Teacher Education, 59*(5), 389–407.

Conference Board of the Mathematical Sciences. (2001). *The mathematical education of teachers.* Providence, RI and Washington, DC: American Mathematical Society and Mathematical Association of America.

Conference Board of the Mathematical Sciences. (2012). *The mathematical education of teachers II.* Providence, RI and Washington, DC: American Mathematical Society and Mathematical Association of America.

Mathematicsonline. (2011, March 18). Area of a circle, how to get the formula [Video file]. Retrieved from https://www.youtube.com/watch?v=YokKp3pwVFc

National Council of Teachers of Mathematics. (2012). CAEP Standards. Retrieved from http://www.nctm.org/Standards-and-Positions/CAEP-Standards/

Limits

Concepts and Applications

Think

When you think about the term *limit*, what thoughts come to mind? Perhaps you think of the speed limit, which indicates a maximum allowable speed on the road. Within insurance policies, there are limits of liability that explain the maximum amount the company will pay. You may have overheard someone say that they are "approaching their limit." These everyday terms may help you, and your future students, understand the mathematical meaning of limits, although the mathematical meaning is more complex and specific. In real life, a limit may be a type of boundary that is approached but never reached; in mathematics, there are cases where the limit can be reached … sometimes in a finite number of steps.

Remember

To prepare for this chapter, you may want to think about the following topics. They serve as prerequisites and primers to the mathematical content in the chapter.

- Identify and describe the pattern in a sequence, such as (a) 1013, 1023, 1033, 1043, 1053, … ; (b) 234, 23.4, 2.34, 0.234, 0.0234, … ; and (c) 1, 1.1, 1.11, 1.111, 1.1111… .
- Perform arithmetic with fractions, such as $\frac{1}{3} + \frac{7}{12} - \frac{5}{48}$ or $2\frac{3}{8} \div \left(-\frac{2}{3}\right)$.
- Write a decimal in expanded notation, as in $47.18 = 4(10) + 7(1) + 1\left(\frac{1}{10}\right) + 8\left(\frac{1}{100}\right)$.
- Understand the circumference as the perimeter of a circle.
- Use technology to create the graph of a function, such as $f(x) = 2x + x^2$ and $g(x) = e^x$.

Connect

The mathematical concept of limit is foundational to other concepts in calculus, and it also underlies several mathematical topics in the middle-grades curriculum. In this chapter, we will apply limits to the contexts of repeating decimals, the perimeter of a circle, and locating a hole in the graph of a function.

Repeating decimals appear in grades 7 and 8 of the Common Core State Standards for Mathematics (CCSS-M) (n.d.). The perimeter of a circle is also addressed in grade 7 in both the CCSS-M and in the Texas Essential Knowledge and Skills (TEKS) (Texas Education Agency, n.d.). The use of limits to find the location of a hole in a graph can be applied to finding decimal approximations for irrational numbers such as π and $\sqrt{2}$. Irrational numbers are introduced in the middle grades as well, explicitly in grade 8.

1.1 Sequences and Series

OBJECTIVES FOR SECTION 1.1: Upon completing this section, you will be able to do the following:

- Identify and create infinite sequences
- Identify and create infinite series
- Create a sequence of partial sums
- Infer whether the value of an infinite series exists

A *sequence* in mathematics is a set of numbers in a particular order. Each number is a *term*.

$$8, 6, 7, 5, 3, 0, 9 \text{ is a sequence. There are seven terms in this sequence.}$$

An *infinite sequence* is a sequence that has an infinite number of terms. The previous sequence is not an infinite sequence. Here are three different infinite sequences.

$$2, 4, 6, 8, 10, 12, \ldots$$

$$\frac{1}{2}, \frac{1}{4}, \frac{1}{8}, \frac{1}{16}, \frac{1}{32}, \ldots$$

$$3, 1, 4, 1, 5, 9, 2, \ldots$$

The terms in a sequence do not have to follow a pattern or rule. If there is a pattern to the terms, we may be able to determine a particular term in a sequence, or even figure out what happens to the sequence as the number of terms gets very large.

A *series* is the sum of the terms in a sequence. For the sequence 8, 6, 7, 5, 3, 0, 9, the series is

$$8 + 6 + 7 + 5 + 3 + 0 + 9. \text{ The } value \text{ of this series is 38.}$$

Like sequences, series may be finite or infinite. Note that a sequence is a set of numbers, but the value of a series is a single number.

Infinite series present us with an interesting problem. While it easy enough to replace the commas with plus signs, as in $2 + 4 + 6 + 8 + 10 + 12 + \ldots$ and $\frac{1}{2} + \frac{1}{4} + \frac{1}{8} + \frac{1}{16} + \frac{1}{32} + \ldots$, how will we determine the value of the series when there are an infinite number of terms to add?

Exercise 1.1: Make an educated guess about the values of these infinite series:

$$2 + 4 + 6 + 8 + 10 + 12 + \ldots =$$

$$\frac{1}{2} + \frac{1}{4} + \frac{1}{8} + \frac{1}{16} + \frac{1}{32} + \cdots =$$

Explain your reasoning.

To tackle the problem of finding the value of an infinite series, we can create a sequence of *partial sums*. We then examine the sequence of partial sums to see if it is approaching a single real number. If the sequence of partial sums approaches a single real number, then that number is the value of the infinite series.

A *partial sum* is the sum of the first few terms of a sequence. We will denote the first partial sum as S_1, the second as S_2, the third as S_3, and so on. S_1 is really just the first term of the sequence. S_2 is the sum of the first two terms, and S_3 is the sum of the first three terms. For example, if we use the sequence 2, 4, 6, 8, 10, 12, ..., we have the following partial sums:

$$S_1 = 2$$
$$S_2 = 2 + 4 = 6$$
$$S_3 = 2 + 4 + 6 = 12$$
$$S_4 = 2 + 4 + 6 + 8 = 20$$
$$S_5 = 2 + 4 + 6 + 8 + 10 = 30$$

This sequence of partial sums appears to increase without bound. That is to say, each partial sum is larger than the previous partial sum, and the difference between consecutive partial sums increases as the number of terms increase. As the number of terms gets very large, it appears that the sequence of partial sums approaches positive infinity. Because positive infinity is not a real number, we say that the value of this infinite series does not exist.

Box 1.1: Possible Pitfalls

If the sequence of partial sums approaches ∞, $-\infty$, or more than one real number, then the value of the series does not exist.

Problem Set 1.1

For each series in problems 1–8, find the first five partial sums.
1. $1 + 2 + 6 + 24 + 120 + 720 + \ldots$
2. $9 + 5 + 1 - 3 - 7 - \ldots$
3. $\frac{1}{2} + \frac{1}{4} + \frac{1}{8} + \frac{1}{16} + \frac{1}{32} + \cdots$

4. $\frac{1}{2} + \frac{3}{4} + \frac{9}{8} + \frac{27}{16} + \frac{81}{32} + \cdots$
5. $6 + 18 + 54 + 162 + \ldots$
6. $7 - 5.6 + 4.48 - 3.584 + 2.8672 - \ldots$
7. $30 + 3 + 0.3 + 0.03 + 0.003 + \ldots$
8. $1 + 0.2 + 0.03 + 0.004 + \ldots$
9. For each of the series in problems 1–8, does the sum appear to exist? Explain.

1.2 Infinite Geometric Series

OBJECTIVES FOR SECTION 1.2: Upon completing this section, you will be able to do the following:

- Identify an infinite geometric series
- Determine the initial term and common ratio of a geometric series
- Use a spreadsheet to construct (part of) an infinite geometric series
- Determine whether the value of an infinite geometric series exists
- Calculate the value of an infinite geometric series (if it exists)

In this section, we will examine a particular type of infinite series called a geometric series. An infinite geometric series has this form:

$$a + ar + ar^2 + ar^3 + ar^4 + \ldots$$

The number a is called the *initial term*, and r is called the common ratio.

Exercise 1.2: Are any of the series shown infinite geometric series? If so, list the values of a and r. If not, explain why.

$2 + 4 + 6 + 8 + 10 + 12 + \ldots$ $\qquad$ $30 + 3 + 0.3 + 0.03 + 0.003 + \ldots$

$\frac{1}{2} + \frac{1}{4} + \frac{1}{8} + \frac{1}{16} + \frac{1}{32} + \cdots$ $\qquad$ $6 + 18 + 54 + 162$

$\frac{1}{2} + \frac{3}{4} + \frac{9}{8} + \frac{27}{16} + \frac{81}{32} + \cdots$ $\qquad$ $7 - 5.6 + 4.48 - 3.584 + 2.8672 - \ldots$

Consider the infinite geometric series $\frac{1}{2} + \frac{3}{10} + \frac{9}{50} + \frac{27}{250} + \cdots$. It should be clear that $a = \frac{1}{2}$. The common ratio is $r = \frac{3}{5}$. One way to see this is that, given a term in the series, we multiply the numerator by 3 and the denominator by 5 to obtain the next term. Another way is to divide a term by the previous term. For an infinite geometric series, this quotient will be the same regardless of the term you choose.

Here are the first five partial sums for this series, written both as fractions and decimals.

$$S_1 = \frac{1}{2} \qquad = \frac{1}{2} \qquad = 0.5$$

$$S_2 = \frac{1}{2} + \frac{3}{10} \qquad = \frac{8}{10} \qquad = 0.8$$

$$S_3 = \frac{1}{2} + \frac{3}{10} + \frac{9}{50} \qquad = \frac{49}{50} \qquad = 0.98$$

$$S_4 = \frac{1}{2} + \frac{3}{10} + \frac{9}{50} + \frac{27}{250} \qquad = \frac{272}{250} \qquad = 1.088$$

$$S_5 = \frac{1}{2} + \frac{3}{10} + \frac{9}{50} + \frac{27}{250} + \frac{81}{1250} \qquad = \frac{1441}{1250} \qquad = 1.1528$$

> **Box 1.2: Possible Pitfalls**
>
> When adding fractions, don't forget to use common denominators.

We can see that the partial sums are increasing, but the amount of increase is decreasing each time. In other words, they are increasing in a way that they may actually approach a real number and not positive infinity. If we continue the sequence of partial sums, we may see them approach the real number that is equal to the value of $\frac{1}{2} + \frac{3}{10} + \frac{9}{50} + \frac{27}{250} + \cdots$.

Exercise 1.3: Here are three infinite geometric series. For each case,

- identify the common ratio,
- find the first five partial sums, and
- tell whether the sum of the infinite series exists.

If the sum exists, estimate the value of the series. If the sum does not exist, explain why not.

$$\frac{1}{2} + \frac{1}{4} + \frac{1}{8} + \frac{1}{16} + \frac{1}{32} + \cdots \qquad 1 + 2 + 4 + 8 + \ldots \qquad 6 - 2 + \frac{2}{3} - \frac{2}{9} + \frac{2}{27} + \cdots$$

We can use a spreadsheet to help find the partial sums for an infinite geometric series. Here is an example for the infinite geometric series $3 + 6 + 12 + 24 + 36 + 72 + \ldots$, with $a = 3$ and $r = 2$.

Open a spreadsheet and type in the information that is shown in Figure 1.1.

In spreadsheets, formulas begin with an equal sign, as you see in cells E2 and F3 (and other cells) in Figure 1.1. The spreadsheet will compute the values for these formulas once you press the return key. You should see the results in Figure 1.2 on the screen when you are finished.

As you may have guessed, the letter-number pairs refer to the cells in the spreadsheet. These are called *relative references*. The added symbol $, as in B$2, creates an *absolute reference*. By using formulas with relative and absolute references, we use the spreadsheet to calculate many partial sums very quickly. We will use a feature called "fill down" to do so.

	A	B	C	D	E	F
1	initial value	common ratio		term	sequence	partial sums
2	3	2		1	=A2	=SUM(E$2:E2)
3				=D2+1	=E2*B$2	=SUM(E$2:E3)

FIGURE 1.1. Formulas for a spreadsheet to make a geometric sequence and series.

	A	B	C	D	E	F
1	initial value	common ratio		term	sequence	partial sums
2	3	2		1	3	3
3				2	6	9

FIGURE 1.2. Using a spreadsheet to make a geometric sequence and series.

One way to "fill down" is to select the group of cells from D2 to F2. Notice a small square in the lower right corner of the selection. Click and drag that square down for several rows. When you release the square, the cells should fill in automatically. Alternatively, you can select a group of cells that is several rows tall and has D2, E2, and F2 in the top row. Use the menus to select Edit > Fill > Down.

Notice from the spreadsheet output that $S_{10} = 3069$ and $S_{25} = 100{,}663{,}293$. It appears that the sequence of partial sums increases to infinity, and therefore the sum of the series does not exist.

Now, we behold the beauty of the spreadsheet: We can use this spreadsheet to examine many infinite geometric series. Modify the initial value and common ratio in cells A2 and B2, and the sequence and partial sums are automatically recomputed. This is awesome!

Exercise 1.4: Use your spreadsheet to find an estimate of the value of each infinite geometric series, if it exists.

$$\frac{1}{2} + \frac{1}{4} + \frac{1}{8} + \frac{1}{16} + \frac{1}{32} + \cdots \qquad 1 + 2 + 4 + 8 \ldots$$

$$6 - 2 + \frac{2}{3} - \frac{2}{9} + \frac{2}{27} + \cdots \qquad \frac{1}{2} + \frac{3}{10} + \frac{9}{50} + \frac{27}{250} + \cdots$$

$$\frac{1}{2} - \frac{3}{4} + \frac{9}{8} - \frac{27}{16} + \frac{81}{32} + \cdots \qquad 7 - 5.6 + 4.48 - 3.584 + 2.8672 - \ldots$$

Exercise 1.5: Reflect on the results you obtained from the spreadsheet in Exercise 1.4. How does the value of a impact the **existence** of the value of the infinite geometric series? How does the value of r impact the **existence** of the value of the infinite geometric series?

We can use some algebra to find a formula for the value of an infinite geometric series, provided that the conditions for a and r are satisfied.

First, I will assume the value of the geometric series exists, and I will call this sum S. Therefore, we have $S = a + ar + ar^2 + ar^3 + ar^4 + \ldots$. Next, I will multiply both sides of this equation by r, which is $Sr = ar + ar^2 + ar^3 + ar^4 + ar^5 + \ldots$. Now, we take the first equation and subtract the second. You should have $S - Sr = a$ because all the other terms on the right disappear through the subtraction. To solve for S, we factor the left-hand side, giving $S(1 - r) = a$. Finally, divide both sides by the quantity $1 - r$ to obtain $S = \frac{a}{1-r}$.

Exercise 1.6: Use this formula on the infinite geometric series to check your work on the spreadsheet. Do you always get the same answers?

Keep in mind that this formula can "take" any value of a and r, but the results may not make sense if the value of the series does not exist. Recall the infinite geometric series with $a = 3$ and $r = 2$ had partial sums that appeared to approach positive infinity. Applying this formula gives us $S = \frac{3}{1-2} = \frac{3}{-1} = -3$, which implies that the sum of an infinite number of positive numbers is a negative number. This makes no sense! The error was that this formula only works when the value of the infinite geometric series exists.

Exercise 1.7: Use your prior work to complete these statements:
The value of an infinite geometric series exists when r is between _____ and _____.
Therefore, if $S = a + ar + ar^2 + ar^3 + ar^4 + \ldots$, then $S = \frac{a}{1-r}$ as long as _____ $< r <$ _____.

Box 1.3: Possible Pitfalls

For infinite geometric series, always check the value of r to see if the formula applies.

Problem Set 1.2

1. Is $1 - \frac{1}{2} + \frac{1}{4} + \frac{1}{8} + \frac{1}{16} - \cdots$ a geometric series? If so, what are a and r? If not, why not?
2. Is $1 + \frac{1}{2} + \frac{1}{3} + \frac{1}{4} + \frac{1}{5} + \cdots$ a geometric series? If so, what are a and r? If not, why not?
3. Is $5 - 10 + 20 - 40 + 80 - \cdots$ a geometric series? If so, what are a and r? If not, why not?
4. Is $2 + \frac{1}{2} + \frac{1}{8} + \frac{1}{32} + \cdots$ a geometric series? If so, what are a and r? If not, why not?
5. Is $-3 + 1 - \frac{1}{3} + \frac{1}{9} + \frac{1}{27} + \cdots$ a geometric series? If so, what are a and r? If not, why not?
6. Is $\frac{1}{10} + \frac{3}{10^2} + \frac{5}{10^3} + \frac{7}{10^4} + \cdots$ a geometric series? If so, what are a and r? If not, why not?
7. Is $\frac{8}{10} + \frac{8}{10^2} + \frac{8}{10^3} + \frac{8}{10^4} + \cdots$ a geometric series? If so, what are a and r? If not, why not?
8. Is $\frac{143}{10^3} + \frac{143}{10^6} + \frac{143}{10^9} + \frac{143}{10^{12}} + \cdots$ a geometric series? If so, what are a and r? If not, why not?
9. Find the value of the series in problem 1. If this is not possible, explain why.
10. Find the value of the series in problem 3. If this is not possible, explain why.

11. Find the value of the series in problem 5. If this is not possible, explain why.
12. Find the value of the series in problem 7. If this is not possible, explain why.
13. Find the value of the series in problem 8. If this is not possible, explain why.
14. Watch the video of a bouncing ping pong ball (katluk, 2008) at https://www.youtube.com/watch?v=A6I3JI4vJv0.

    Let's suppose that a ping pong ball is dropped from 6 inches above a table. With each bounce, it travels up 80% of the previous height.

    a. Complete the table for the vertical distances of the first five bounces.

Bounce number	Distance down	Distance up
1	6	
2		
3		
4		
5		

    b. How much distance (vertically) has the ping pong ball traveled after five bounces?
    c. How do infinite geometric series apply to this problem?
    d. If there was no friction, the ping pong ball would theoretically bounce an infinite number of times. How far would the ping pong ball travel vertically in that case?

1.3 Repeating Decimals

OBJECTIVES FOR SECTION 1.3: Upon completing this section, you will be able to do the following:

- Write repeating decimals in a way that is (or contains) an infinite geometric series
- Use infinite geometric series to write a repeating decimal as a fraction

Students in grades 4–8 continue to develop an understanding of rational numbers, including fractions and decimal representations. In particular, students may be introduced to repeating decimals during this time. Repeating decimals can be written in a way that contains an infinite geometric series. Some are straightforward when written in expanded notation and others require a little more finesse.

Box 1.4: Teaching Tips

Middle-school students may not attend to the bar over the digits. They may also think 0.3 is the same as $0.\overline{3}$. Ask them to subtract one from the other. The difference isn't zero, so the numbers are different.

Let's begin with a straightforward example: $0.4444\ldots = 0.\overline{4}$.

Exercise 1.8: I claim $0.4444\ldots = 0.\overline{4} = \dfrac{4}{10} + \dfrac{4}{10^2} + \dfrac{4}{10^3} + \dfrac{4}{10^4} + \cdots$

a. Verify that this repeating decimal can be written as the given infinite geometric series.
b. What are a and r?
c. Does the value of r allow us to use the formula $S = \dfrac{a}{1-r}$?
d. Use this formula to write the repeating decimal as a fraction.

Now, let's think about another example, $0.143143143\ldots = 0.\overline{143}$.

$$0.143143143\ldots = 0.\overline{143} = \dfrac{1}{10} + \dfrac{4}{10^2} + \dfrac{3}{10^3} + \dfrac{1}{10^4} + \dfrac{4}{10^5} + \dfrac{3}{10^6} + \cdots$$
$$= \dfrac{143}{10^3} + \dfrac{143}{10^6} + \dfrac{143}{10^9} + \dfrac{143}{10^{12}} + \cdots$$

Exercise 1.9: Two infinite series for $0.143143\ldots$ are provided (see previous equations).

a. Verify that they are equal to the repeating decimal.
b. Only one series is an infinite geometric series. Which one is it?
c. Are we allowed to use the formula $S = \dfrac{a}{1-r}$ for this repeating decimal?
d. Find the sum of the series as a fraction or explain why it can't be done.

Finally, consider the repeating decimal $0.7181818\ldots = 0.7\overline{18}$. As in the last case, I will write two different infinite series to represent the repeating decimal.

$$0.7181818\ldots = 0.7\overline{18} = \dfrac{7}{10} + \dfrac{1}{10^2} + \dfrac{8}{10^3} + \dfrac{1}{10^4} + \cdots$$
$$= \dfrac{7}{10} + \dfrac{18}{10^3} + \dfrac{18}{10^5} + \dfrac{18}{10^7} + \cdots$$

Exercise 1.10: Two infinite series for $0.7181818\ldots$ are provided (see previous equations).

a. Verify that they are equal to the repeating decimal.
b. Notice that neither of these is an infinite geometric series. Why not?
c. The second representation contains an infinite geometric series. For this part of the series, answer the following questions:
   i. What are a and r?
   ii. Does the value of r allow us to use the formula $S = \dfrac{a}{1-r}$?
   iii. Use this formula to find write the repeating decimal as a fraction.
   iv. Explain how you incorporated the $\dfrac{7}{10}$ in your solution process.

Problem Set 1.3

1. Write the repeating decimal $0.\overline{8}$ as an infinite geometric series.
2. Write $0.\overline{8}$ as a fraction. Show work that displays how you used calculus ideas.
3. Write the repeating decimal $0.\overline{34}$ as an infinite geometric series.
4. Write $0.\overline{34}$ as a fraction. Show work that displays how you used calculus ideas.
5. Write the repeating decimal $0.\overline{281}$ as an infinite geometric series.
6. Write $0.\overline{281}$ as a fraction. Show work that displays how you used calculus ideas.
7. Write the repeating decimal $0.\overline{9}$ as an infinite geometric series.
8. Write $0.\overline{9}$ as a fraction. Show work that displays how you used calculus ideas.
9. Write the repeating decimal $0.8\overline{25}$ in a form that is (or contains) an infinite geometric series.
10. Write $0.8\overline{25}$ as a fraction. Show work that displays how you used calculus ideas.
11. Write the repeating decimal $5.5\overline{6}$ in a form that is (or contains) an infinite geometric series.
12. Write $5.5\overline{6}$ as a fraction. Show work that displays how you used calculus ideas.
13. Write the repeating decimal $0.12\overline{08}$ in a form that is (or contains) an infinite geometric series.
14. Write $0.12\overline{08}$ as a fraction. Show work that displays how you used calculus ideas.
15. Two students, Simon and Maryam, are discussing decimals. Simon says that there are three kinds of decimals: (a) terminating, like 0.25, (b) repeating, like $0.\overline{3}$, and (c) decimals that go on forever and never repeat, like π. Maryam disagrees with Simon, and says that what Simon calls a terminating decimal is really a kind of repeating decimal. She says that $0.25 = 0.25\overline{0}$, with a repeating 0. As a teacher, how do you respond to Maryam and Simon?

1.4 Perimeter of a Circle

OBJECTIVES FOR SECTION 1.4: Upon completing this section, you will be able to do the following:

- Read, interpret, and write limit notations
- Understand the approach Archimedes used to determine the perimeter of a circle
- Use approximations to determine the limit of a sequence to a specified level of accuracy.

1.4.1 Limit Notation

In the sequence $\frac{1}{2}, \frac{1}{4}, \frac{1}{8}, \frac{1}{16}, \frac{1}{32}, \cdots, \frac{1}{2^n}, \cdots$, we see that the terms get smaller and smaller, and approach zero as the number of terms increases to infinity. We may use *limit notation* to express this idea, as shown here:

$$\lim_{n \to \infty}\left(\frac{1}{2^n}\right) = 0$$

We read this as, "The limit of $\frac{1}{2^n}$ as n approaches infinity is 0."

In section 1.2, we saw that the value of the series based on this sequence was 1. That is,

$$\frac{1}{2} + \frac{1}{4} + \frac{1}{8} + \frac{1}{16} + \frac{1}{32} + \cdots = 1.$$

We can arrive at this idea by examining the sequence of partial sums.

$S_1, S_2, S_3, S_4, S_5, \cdots = \frac{1}{2}, \frac{3}{4}, \frac{7}{8}, \frac{15}{16}, \frac{31}{32}, \cdots$. Using limit notation, we write $\lim_{n \to \infty}(S_n) = 1$.

1.4.2 Measuring Around a Circle

The idea of limits is complex but also extremely helpful. Limits have many uses in calculus and have been used to solve problems that were once very difficult in the past. One such problem was finding the perimeter of a circle. Nowadays, we know a formula. But where did the formula come from? How do we know it is correct? Ideas from calculus help show that the formulas used in middle school are, in fact, correct and have a rationale.

> **Box 1.5: Teaching Tips**
>
> In the middle-school classroom, you can use a measuring tape or string to measure the perimeter and diameters of several different circles.

About 2,300 years ago, it was much easier to measure the lengths of straight line segments than it was to measure curves. This is still true today. To tackle the problem of measuring the perimeter of a circle, we will use the approach of Archimedes of Syracuse (287–212 BC).

Archimedes used regular polygons to approximate circles. These polygons were made of straight line segments, so it was much easier to find their perimeters. He examined both *inscribed* and *circumscribed* polygons. Inscribed polygons are inside the circle, and their vertices are located on the circle. Circumscribed polygons are outside of the circle, and each side of the polygon touches the circle exactly once.

We will use GeoGebra to explore this problem and attempt to understand how Archimedes accomplished this feat. Go to https://ggbm.at/BN8D2duN, and open the worksheet "1.4 Perimeter of a Circle."

This file uses a circle centered at the origin, with a diameter of 1 unit, and n is the number of sides of the inscribed and circumscribed regular polygons. Use the following exercises to guide you through an investigation. Be sure to record your answers in writing.

Exercise 1.11: First, examine the inscribed polygons.

a. How do the perimeters of the inscribed polygons compare to the perimeter of the circle? That is, are they smaller or larger?
b. What happens to the inscribed polygons as the number of sides increases?

c. Why is the following statement true?

$$\lim_{n \to \infty} (\text{perimeter of inscribed polygon}) = \text{perimeter of circle}$$

d. Write down a sequence of the perimeters of the first six inscribed polygons.
e. How would you describe this sequence? Is it increasing, decreasing, or staying constant?

Exercise 1.12: Now, examine the circumscribed polygons.

a. How do the perimeters of the circumscribed polygons compare to the perimeter of the circle?
b. What happens to the circumscribed polygons as the number of sides increases?
c. Why is the following statement true?

$$\lim_{n \to \infty} (\text{perimeter of inscribed polygon}) = \text{perimeter of circle}$$

d. Write down a sequence of the perimeters of the first six circumscribed polygons.
e. How would you describe this sequence? Is it increasing, decreasing, or staying constant?

Exercise 1.13: Now, consider both inscribed and circumscribed polygons.

a. Using inscribed and circumscribed octagons ($n = 8$), we can say that the perimeter of this circle is between which two values? _____ and _____. Our estimate (using octagons) is _____, correct to the _____ place.
b. Using inscribed and circumscribed henahectagons ($n = 100$), we can say that the perimeter of this circle is between which two values? _____ and _____. Our estimate (using henahectagons) is _____, correct to the _____ place.

Archimedes used these same ideas, but obviously not the same technology, to estimate the perimeter of a circle and the value of π. He began with a regular hexagon. He used formulas to compute the new perimeters based on doubling the number of sides of the polygon. He began with regular hexagons ($n = 6$) and doubled the number of sides until he reached enneacontakaihexagons ($n = 96$).

Problem Set 1.4

Use the GeoGebra applet "1.4 Perimeter of a Circle" at https://ggbm.at/BN8D2duN to investigate the perimeter of a circle, following Archimedes's method of doubling the number of sides. Note that this file has a circle centered at the origin with a **diameter of 1 unit.**

Regular polygons with n sides have been inscribed in and circumscribed about the circle. With technology, it is not difficult to use more sides that Archimedes examined.

1. Complete the chart in Table 1.1.
2. Write a brief paragraph to describe the results in the table from problem 1.
3. The number π is defined as the ratio of the circumference of a circle and its diameter. Because we used a diameter of 1 unit, we can use these sequences to estimate the value of π. Use GeoGebra to complete the Table 1.2 about the accuracy of our estimate for π.

TABLE 1.1   Archimedes's Approach to Finding the Perimeter of a Circle with Diameter = 1

Number of sides, n	Perimeter of inscribed polygon	Perimeter of circumscribed polygon
6		
12		
24		
48		
96		
192		
384		
Limit as $n \to \infty$		

TABLE 1.2   Approximating the Value of π

Approximation of π	Accurate to this decimal place	Minimum n to achieve this level of accuracy
3	ones	
	tenths	
3.14		
3.141	thousandths	160
	ten-thousandths	

1.5 Location of a Hole in a Graph

OBJECTIVES FOR SECTION 1.5: Upon completing this section, you will be able to do the following:

- Use technology to investigate functions with discontinuities
- Apply limits to approximate function values to specified levels of accuracy
- Determine the location of a hole in a graph of a function

The function $f(x) = \dfrac{10\sqrt{x+7} - 30}{x-2}$ has a hole[1] in it. In this section, we will approximate the location of the hole using Desmos, a free graphing calculator located at http://www.desmos.com.

Exercise 1.14: Let's find the coordinates of where this hole in the function should be located.

a. Using your knowledge about the domains of functions, which x-values from the real numbers are not in the domain? This will help us find the x-value of the hole.
b. Launch the calculator on the website and type in the formula for this function.
c. Does the graph of the function show the hole?
d. When you click the graph of the function, you should see a point highlighted and the coordinates $(x, f(x))$. Move the highlighted point to the place where you think the hole should be. What does Desmos give for the coordinates at this point?
e. Add a table and name the columns x and $f(x)$. Does the table indicate that there is a hole?

We would like to find the y-value of the hole. Since directly inputting the x-value of 2 doesn't work, we will use the technique of choosing x-values close to the x-value of the hole. This will give us a sequence of y-values. Try this with two different sequences of x-values: one that approaches the x-value of the hole from the left and another that approaches it from the right.

Approaching 2 from the left

x	$f(x)$
1	
1.9	
1.99	
1.999	

Approaching 2 from the right

x	$f(x)$
3	
2.1	
2.01	
2.001	

---

[1] I wish to acknowledge Dr. Jason Martin (University of Central Arkansas) for showing me this idea during the CLEAR Calculus workshop presented at Sam Houston State University.

Exercise 1.15: Use these tables to complete the following statements:

a. The coordinates of the hole are approximately (2, _____).
b. This estimate is correct to the _____ decimal place. Explain how you know.
c. $\lim_{x \to 2}(f(x)) = $ _____, correct to _____ decimal places.

Problem Set 1.5

For problems 1–3, use some sort of technology to find the location of the hole in each of the functions. Do this by making approximations. Also, complete the limit statements, with your approximations correct to five decimal places.

1. The graph of $f(x) = \dfrac{\sqrt{x+1} - 2}{x - 3}$ has a hole at (_____ , _____).

   $\lim_{x \to ___}(f(x)) = $ _____ , correct to 5 decimal places.

2. The graph of $g(x) = \dfrac{e^x - e^2}{x - 2}$ has a hole at (_____ , _____).

   $\lim_{x \to ___}(g(x)) = $ _____ , correct to 5 decimal places.

3. The graph of $h(x) = \left(\dfrac{x+2}{2}\right)^{\frac{1}{x}}$ has a hole at (_____ , _____).

   $\lim_{x \to ___}(h(x)) = $ _____ , correct to 5 decimal places.

4. Ashton and Brady are discussing the function $k(x) = \dfrac{3 \cdot \ln(x)}{x - 1}$.

   a. Ashton says the graph of the function has a hole at (1, 3). Do you agree with Ashton? Explain why or why not.

   b. Brady made the following tables.

x	k(x)	x	k(x)
0.9	3.1608155	1.1	2.8593054
0.99	3.0151008	1.01	2.9850993
0.999	3.001501	1.001	2.998501
0.9999	3.00015	1.0001	2.99985
0.99999	3.000015	1.00001	2.999985

   He says he cannot find an approximation of $\lim_{x \to 1}(k(x))$ correct to five decimal places because these approximations never agree. How do you respond to Brady's concern?

References

Common Core State Standards Initiative. (n.d.). Retrieved from http://www.corestandards.org

katluk. (2008, October 9). Animation reference: Bouncing ball (ping pong) [Video file]. Retrieved from https://www.youtube.com/watch?v=A6I3JI4vJv0

Texas Education Agency. (n.d.). Texas education knowledge and skills. Retrieved from https://tea.texas.gov/curriculum/teks/

Credits

Fig. 1.1: Copyright © by Google.
Fig. 1.2: Copyright © by Google.

Derivatives

Concepts

Think

When you think about the phrase *rate of change*, what comes to mind? Perhaps you think of the formula *distance = rate × time*, or a linear function that has a constant rate of change. You may also think about graphs of lines, in which the slope is a ratio of the changes in two values—one on the horizontal axis, and the other on the vertical axis. This has been referred to as *rise over run* and is symbolized as $\frac{\Delta y}{\Delta x}$, meaning a change in y over a change in x. In this chapter, we will examine rates of change for linear and non-linear functions. For non-linear functions, the rate of change is not a constant. We will use the broader term *derivative* to describe these rates of change, because they are derived (or taken) from the function.

Remember

To prepare for this chapter, you may want to think about the following topics. They serve as prerequisites and primers to the mathematical content in the chapter.

- Determine the average rate of change of a vehicle, given the distance traveled over a period of time—for example, a car that travels 6 miles in 15 minutes.
- Calculate the slope of a line given two points (such as (1,5) and (4,14)) or an equation (such as $y = 3x + 2$, or $f(x) = 5$).
- Recall the appearance of a line with a slope that is negative, positive, zero, or undefined.
- Evaluate functions with specific inputs. That is, if $f(x) = x^2 + x - 1$, find $f(4)$, $f(a)$, $f(a + 4)$, and $f(4 - a)$.

Connect

In the middle grades, the focus on rate of change lies primarily in the realm of linear functions and proportional relationships. We will begin our discussion within the familiar context of speed as the rate of change of distance with respect to time, and then expand into other areas where the rate of change is not constant.

In grade 6, there is a heavy emphasis on ratios and rates, both in the Texas Essential Knowledge and Skills (TEKS) (n.d.) and the Common Core State Standards for Mathematics (CCSS-M) (n.d.). The ideas are typically introduced through proportional relationships. Through grades 7 and 8, these ideas are further developed and extended to linear models, with an emphasis on linear functions and slope in grade 8. Throughout the middle grades, connections are made between graphical, numerical, tabular, and verbal descriptions of situations.

2.1 How Do We Measure Speed?

OBJECTIVES FOR SECTION 2.1: Upon completing this section, you will be able to do the following:

- Calculate the average velocity of an object over an interval
- Use sequences of average velocities to approximate instantaneous velocity
- Interpret and construct graphs of functions with respect to average and instantaneous velocity.

Watch Tony Schumacher set a speed record in 2005 (LitbyNoonProductions, 2012). The video is located at https://www.youtube.com/embed/rYNeGbOwvnY?start=120&end=155. Pay careful attention to the part from 2:00 to 2:35. According to the announcer and the scoreboard, Mr. Schumacher traveled $\frac{1}{4}$ mile in 4.446 seconds. His record-breaking speed was 337.58 miles per hour ... or was it?

Exercise 2.1: I do not contest the fact that Tony Schumacher was driving fast.

    a. Determine for yourself whether the reported speed of 337.58 miles per hour is correct.
    b. If you get a different result than the scoreboard, why is this so?

Exercise 2.2: Figure 2.1 contains three different distance vs. time graphs for this scenario.

    a. Which of the graphs in Figure 2.1 best describes that of Mr. Schumacher's trip?
    b. Why did you choose this graph?
    c. What do the other two graphs indicate about the speed of the vehicle?

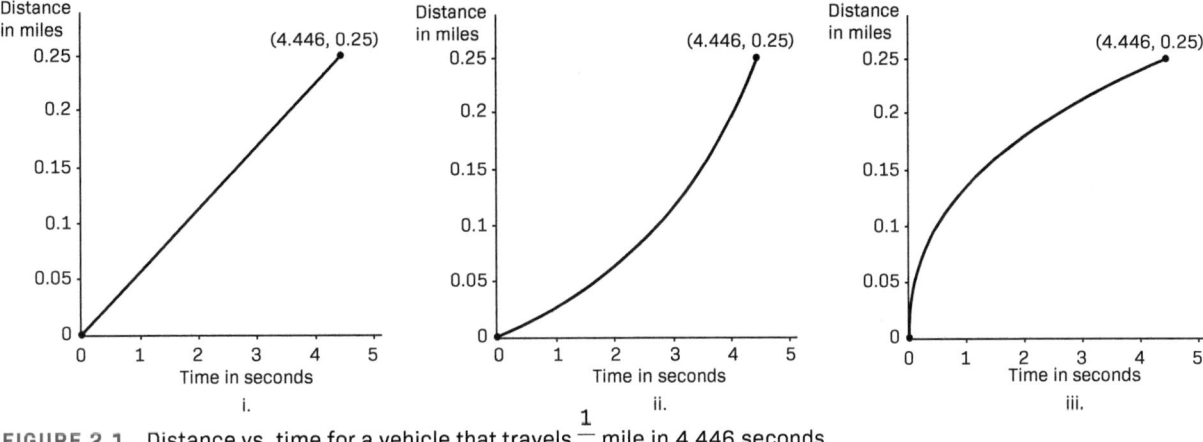

FIGURE 2.1. Distance vs. time for a vehicle that travels $\frac{1}{4}$ mile in 4.446 seconds.

If an object moves at a constant speed during a given time interval, then the graph of distance vs. time will be a straight line. The speed of the object is the slope of the line. It is more accurate to say that the velocity of the object (which tells both magnitude and direction) is the slope of the line. If the velocity is not constant during a given time interval, then the graph of the distance vs. time will not be a straight line. However, we can still use the ideas of slope to consider the velocity of the object at particular points in time.

For an example of an object that changes velocities, watch the slow-motion video (Weierich, 2013) of a person jumping at https://www.youtube.com/watch?v=hhICoM40_ms. Let's focus on his vertical velocity and agree that when the person is moving up, the velocity is positive. Similarly, when he is moving down, his velocity is negative.

Exercise 2.3: Was the person's velocity ever zero? If so, when?

Now watch Dan Meyer's basketball shot (Meyer, 2000) at http://vimeo.com/16832687. The photography effect, replicated in Figure 2.2, allows us to see the ball in 27 positions along its trajectory. The camera recorded the position of the ball 15 times every second.

Exercise 2.4: Focus on how the vertical velocity changes in Figure 2.2.

a. Identify some places when the vertical velocity of the ball was positive.
b. When was the vertical velocity the most negative?
c. Was the vertical velocity ever zero?

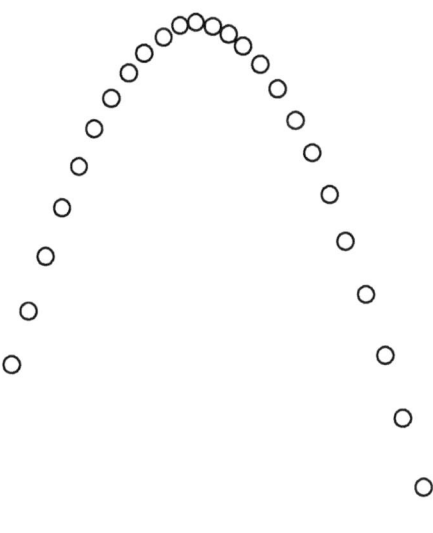

FIGURE 2.2. The path of a basketball.

CHAPTER 2 Derivatives | 25

> **Box 2.1: Possible Pitfalls**
>
> The basketball is moving both horizontally and vertically. Here, let's focus on the up and down movement.

To calculate the *average velocity* of an object over a time interval, we divide the change in position by the length of time of the interval. That is, if $s(t)$ is a function that describes the position of an object at time t, then over the interval $a \leq t \leq b$, the average velocity of the object is equal to $\dfrac{\text{change in position}}{\text{change in time}} = \dfrac{s(b) - s(a)}{b - a}$.

Let's apply this formula in the following example. Suppose a potato is thrown high into the air. Table 2.1 shows the height of the potato ($s(t)$, measured in feet) over a 6-second interval of time (t). The height of the potato was recorded every second.

TABLE 2.1 The Height of a Thrown Potato

t (SEC)	0	1	2	3	4	5	6
s(t) (FEET)	6	90	142	162	150	106	30

We can find the average velocity of the potato for the first 2 seconds by dividing the change in position from 0 seconds to 2 seconds by the length of the 2-second interval.

$$\text{average velocity on } 0 \leq t \leq 2 = \frac{142 \text{ feet} - 6 \text{ feet}}{2 \text{ sec} - 0 \text{ sec}} = \frac{136 \text{ feet}}{2 \text{ sec}} = 68 \frac{\text{feet}}{\text{sec}}$$

Note that this average velocity is positive, meaning that the potato is moving up during this time.

Exercise 2.5: Use Table 2.1.

a. Calculate the average velocity over the interval from 2 seconds to 3 seconds.
b. Will the average velocity for the interval $0 \leq t \leq 5$ be positive or negative? How do you know?
c. Calculate the average velocity for the interval $3 \leq t \leq 6$.

Let's examine the graph of an object that is at a position $s(a)$ at time a, and $s(b)$ at time b, as shown in Figure 2.3. There are many, many curves that could connect these two points. Regardless of the curve, the average velocity will still be calculated in the same way:

$$\text{average velocity over interval } a \leq t \leq b = \frac{s(b) - s(a)}{b - a}$$

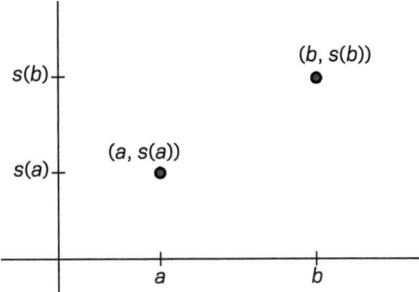

FIGURE 2.3. Two points denoting position s(t) at t = a and t = b.

We have seen this formula before! It is the formula for the slope of a line. Notice in the graph in Figure 2.4 that the slope of the line segment between the two points is the average velocity over the time interval from a to b.

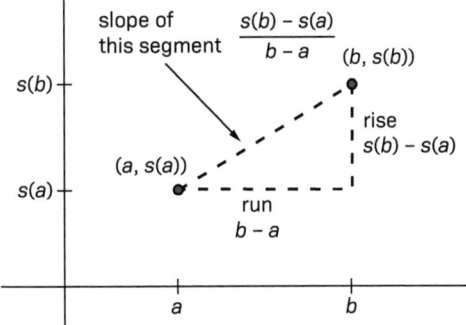

FIGURE 2.4. Average rate of change of s(t) on the interval $a \leq t \leq b$.

Let's practice this with another context: riding a bicycle. Consider the graph of Craig's bicycle ride in Figure 2.5. The distance d is measured in miles from his house, and d is a function of time t.

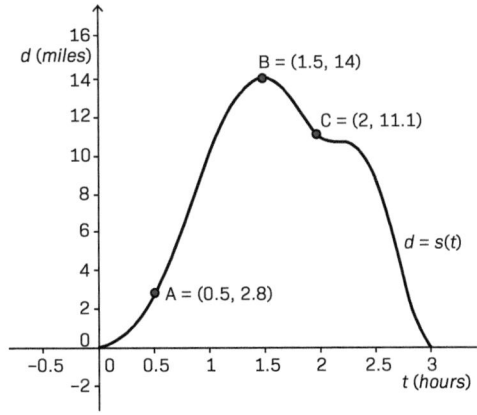

FIGURE 2.5. Craig's bicycle ride.

CHAPTER 2 Derivatives | 27

Exercise 2.6: Are the following positive, negative, or zero?

   I. Craig's average velocity from 0 hours to 1.5 hours
   II. Craig's average velocity from 0.5 hours to 1.5 hours
   III. Craig's average velocity from 0 hours to 2 hours
   IV. Craig's average velocity from 0 hours to 3 hours
   V. Craig's average velocity from 1.5 hours to 2 hours

Order the average velocities described from least (that is, most negative) to greatest.

The *instantaneous velocity* of an object is the velocity of an object at a particular time. It is difficult to calculate the velocity for a particular instant directly with the average rate of change formula, because the times are the same and we obtain zero in the denominator. Instead, we will find the average velocity over a small time interval and then look at a sequence of average velocities as the width of the interval approaches zero.

Using the example of Craig's bicycle ride (Figure 2.5), we could approximate his instantaneous velocity at $a = 0.5$ hr by considering the sequence of average velocities over these intervals.

TABLE 2.2 Intervals That Begin With $a = 0.5$ hr and Have Decreasing Widths

INTERVAL (a TO $a + h$)	0.5 to 1.5	0.5 to 1.0	0.5 to 0.75	0.5 to 0.6	0.5 to 0.55	0.5 to 0.51
WIDTH OF INTERVAL (h)	1.0	0.5	0.25	0.01	0.05	0.01

We could also examine the sequence of average velocities from intervals[1] that end at $a = 0.5$ hr.

TABLE 2.3 Intervals That End with $a = 0.5$ hr and Have Decreasing Widths

INTERVAL ($a - h$ TO a)	0.0 to 0.5	0.25 to 0.5	0.4 to 0.5	0.45 to 0.5	0.49 to 0.5
WIDTH OF INTERVAL (h)	0.5	0.25	0.1	0.05	0.01

At this point, we don't have enough information to calculate some of these average velocities, but we could estimate them with the graph in Figure 2.5.

More formally, if $s(t)$ gives the position of an object at time t, then over an interval from a to $a + h$ we have

$$\text{intanteneous velocity at time } a = \lim_{h \to 0} \left(\frac{s(a+h) - s(a)}{(a+h) - a} \right)$$

$$= \lim_{h \to 0} \left(\frac{s(a+h) - s(a)}{h} \right)$$

---

[1] To think about the width of an interval as a positive value, we are considering $t = a - h$ to $t = a$. One could also consider $t = a + h$ to $t = a$, as in Table 2.2, using values of h that are negative.

In this way, the instantaneous velocity can be viewed as the *slope of the curve at a point*.

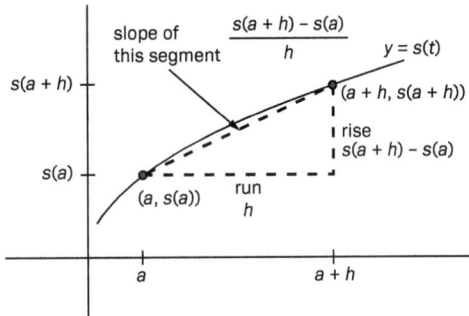

FIGURE 2.6. As *h* approaches 0, we can estimate the instantaneous velocity at $t = a$.

Exercise 2.7: Use the graph of Craig's three-hour bicycle ride, shown here.

a. Are the following positive, negative, or zero?
   i. Craig's instantaneous velocity at 0.5 hours
   ii. Craig's instantaneous velocity at 1.5 hours
   iii. Craig's instantaneous velocity at 2 hours
b. Identify another point (besides A, B, or C) where Craig's instantaneous velocity is positive.
c. Identify another point (besides A, B, or C) where Craig's instantaneous velocity is negative.
d. Identify another point (besides A, B, or C) where Craig's instantaneous velocity is zero.

Support your answers using the idea of slope.

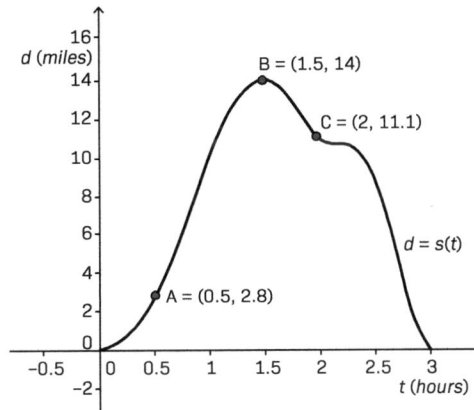

Problem Set 2.1

1. Consider the following scenarios:
    a. A car is driven at a constant speed.
    b. A car is driven at an increasing speed.
    c. A car starts at a slow constant speed, then rapidly accelerates and reaches a fast constant speed.
    d. A car starts at a high speed, and its speed then decreases slowly.

   For each scenario, sketch a graph of the distance the car has traveled as a function of time.

2. A model rocket was launched vertically. Table 2.4 shows the height of a rocket over a six-second interval of time.
    a. Compute the average velocity of the rocket over the interval $4 \leq t \leq 5$.
    b. What is the significance of the sign of your answer in part a.?
    c. Compute the average velocity of the rocket over the interval $1 \leq t \leq 3$.

TABLE 2.4 The Height of a Model Rocket

t (SEC)	0	1	2	3	4	5	6
y (FEET)	3	87	139	159	147	103	27

3. Use the graph of the function shown in Figure 2.7 to answer the following questions:
    a. At what labeled points is the slope of the graph positive?
    b. At what labeled points is the slope of the graph negative?
    c. At which labeled point does the graph have the greatest (i.e., most positive) slope?
    d. At which labeled point does the graph have the least slope (i.e., negative and with the largest magnitude)?

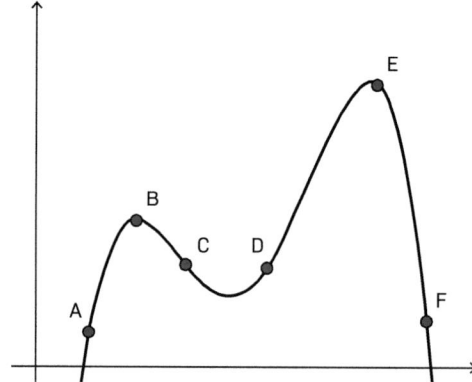

FIGURE 2.7. Graph for problem 3.

4. Use the graph $y = f(x)$ shown in Figure 2.8 to arrange the following numbers from least to greatest.
   a. The slope of the graph at point A
   b. The slope of the graph at point B
   c. The slope of the graph at point C
   d. The slope of the line segment AB
   e. The number 0
   f. The number 1

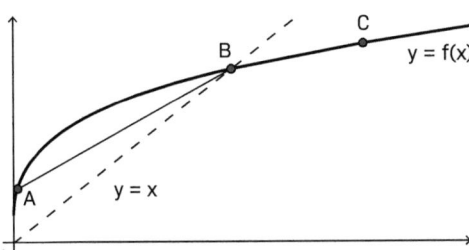

FIGURE 2.8. Graph for problem 4.

5. Suppose a particle is moving at a varying velocity along a straight line and that $s = f(t)$ represents the distance of the particle from a point as a function of time t.
   a. Sketch a possible graph for f so that the average velocity of the particle between $t = 0$ and $t = 3$ is positive.
   b. Sketch a possible graph for f so that the instantaneous velocity at $t = 7$ is 0 and the instantaneous velocity at $t = 8$ is negative.
   c. Sketch a possible graph for f so that the average velocity of the particle between $t = 2$ and $t = 6$ is the same as the instantaneous velocity at $t = 5$.
   d. Sketch a possible graph for f so that the average velocity of the particle between $t = 2$ and $t = 6$ is positive and the instantaneous velocity at $t = 5$ is negative.

6. Estimate $\lim\limits_{h \to 0} \left(\dfrac{(2+h)^3 - 8}{h} \right)$ by substituting smaller and smaller values for h. Make your estimate correct to two decimal places.

7. Estimate $\lim\limits_{h \to 0} \left(\dfrac{\sqrt{4+h} - 2}{h} \right)$ by substituting smaller and smaller values for h. Make your estimate correct to two decimal places.

8. Estimate $\lim\limits_{h \to 0} \left(\dfrac{e^{1+h} - e}{h} \right)$ by substituting smaller and smaller values for h. Make your estimate correct to two decimal places.

2.2 The Derivative at a Point

OBJECTIVES FOR SECTION 2.2: Upon completing this section, you will be able to do the following:

- Approximate the derivative of a function at a point to specified levels of accuracy
- Interpret graphs with respect to the derivative of a function at a point
- Use technology and algebra to determine the derivative of a function at a point
- Write an equation for the line tangent to a curve at a given point.

The ideas of average velocity and instantaneous velocity can be applied in other situations. If f is a function of x, then we may discuss the average rate of change of f and the instantaneous rate of change of f.

> **Box 2.2: Teaching Tips**
>
> Teachers need to know a wide variety of situations where rates of change apply. One is speed, but there are many more described in this chapter, such as the rate of change in the area of a circle with respect to the length of the radius, or the rate of change in the production cost of ice cream with respect to the quantity produced.

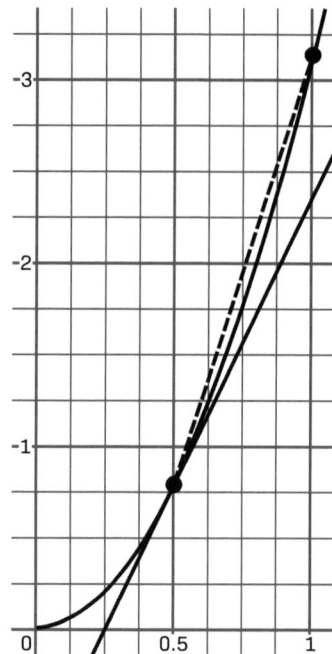

FIGURE 2.9. Estimating the instantaneous rate of change at $r = 0.5$.

For example, consider what happens when a leaky pen creates a circular ink stain that grows larger and larger. We could say that the area of the stain is a function of the radius of the circle, $A = f(r)$. When the radius is 0.5 cm, the area is approximately 0.79 square cm. When the radius is 1 cm, the area is approximately 3.14 square cm. Therefore, we can calculate that the average rate of change of the area over the interval from $r = 0.5$ to $r = 1$ as follows:

$$\frac{f(1)-f(.5)}{1-.05} = \frac{3.14-0.79}{1-0.5} = \frac{2.35}{0.5} = 4.7$$

This means the slope of the dashed line segment in figure 2.9 is 4.7.

Examine the graph of $A = f(r) = \pi r^2$ in figure 2.9 and notice that the line tangent to the function at $r = 0.5$ has a positive slope. This shows that the instantaneous rate of change for the function at $r = 0.5$ is positive.

By comparing the slopes of this line and the dashed segment that connects the points on the graph where $r = 0.5$ and $r = 1$, we see the instantaneous rate of change at $r = 0.5$ is less than the average rate of change over the interval $0.5 \leq r \leq 1$. Therefore, we can say that the average rate of change at $r = 0.5$ is between 0 and 4.7.

What about the line that is tangent to the parabola at $r = 0.5$, the solid line shown in figure 2.9? We only know one point on the line, so how can we find the slope? We can obtain a better estimate for the slope of the tangent line using the grid. It appears that this line passes through the point (0.25, 0). The tangent line appears[2] to have a rise of a little more than 0.75 of a unit and a run of 0.25 unit, so its slope is a little more than $\frac{0.75}{0.25} = 3$. This agrees with our earlier results that the instantaneous rate of change is positive and less than 4.7.

In general, we refer to the instantaneous rate of change of a function as the *derivative* of the function. If the function is named f, then the derivative of the function is denoted as f', pronounced "f prime." Using the circular ink stain example, we may say that $f'(0.5)$ is approximately 3.

It may be helpful to consider that all these phrases essentially mean the same thing:

- The instantaneous rate of change of f at the point where $x = a$
- The slope of the curve $y = f(x)$ at $x = a$
- The slope of the line tangent to $y = f(x)$ at the point where $x = a$
- The derivative of f at $x = a$
- $f'(a)$

Box 2.3: Possible Pitfalls

If you focus on only one meaning of the derivative, you will miss a wide variety of applications and limit your potential for understanding.

Definition 2.2.1: Derivative at a Point

The derivative of a function $f(x)$ at the point where $x = a$ is defined as
$$f'(a) = \lim_{h \to 0} \left(\frac{f(a+h) - f(a)}{h} \right).$$

Recall the item from problem set 2.1 that asked for an approximation of $\lim_{h \to 0} \left(\frac{(2+h)^3 - 8}{h} \right)$. Notice that this problem mirrors the definition of the derivative where the function is $f(x) = x^3$ and $a = 2$. You should have found that this approaches 12, and therefore $f'(2) = 12$. Furthermore, we can say that the slope of the curve $y = x^3$ at the point (2,8) is 12. We can verify this by examining a graph and see that the line just touches the function in one point.

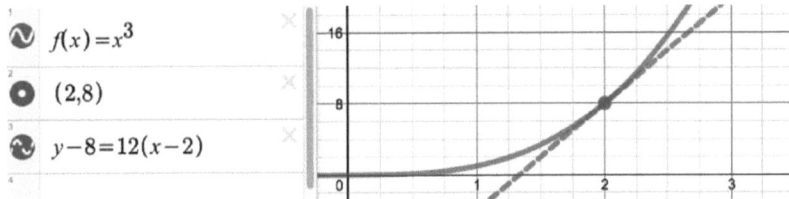

In the same way, your estimate for $\lim_{h \to 0} \left(\frac{\sqrt{4+h} - 2}{h} \right)$ would be $k'(4)$, the slope of a line tangent to $k(x) = \sqrt{x}$ at the point where $x = 4$. Finally, your estimate for $\lim_{h \to 0} \left(\frac{e^{1+h} - e}{h} \right)$ would be $g'(1)$, the slope of a line tangent to $g(x) = e^x$ at the point where $x = 1$.

---

[2] You may obtain a different approximation, but it should be reasonably close to the estimate here. Another possible estimate would use a rise of 1.25 and a run of 0.375, for a slope of $\frac{1.25}{0.375} = 3\frac{1}{3}$.

Exercise 2.8: We can use a graph (with grid lines) to estimate the value of the derivative at a point. Consider the graph of the parabola $f(x) = x^2$, shown in figure 2.10.

a. Plot the point $(1, f(1))$ on the graph. Use a ruler to sketch a line that is tangent to the parabola at that point.
b. Is $f'(1)$ positive, negative, or zero?
c. Use a ruler and the grid to estimate the value of $f'(1)$.
d. Repeat this process to estimate the following:
   i. $f'(0)$
   ii. $f'(-0.5)$

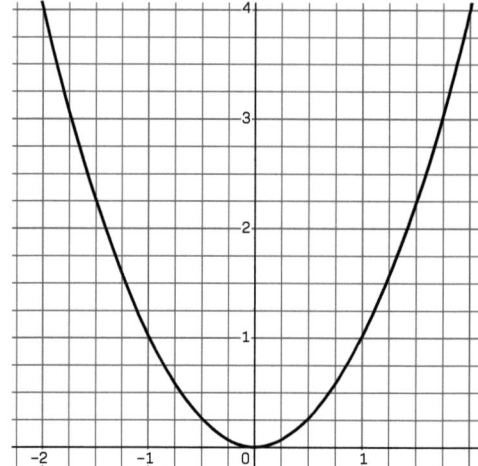

FIGURE 2.10. Graph of a parabola.

We can examine the slope of a curve at a point using Desmos. The set-up is shown in figure 2.11, and includes instructions for the following:

- Plotting the function (line 1)
- Plotting a movable point on the function (lines 2 and 4)
- Plotting a second point on the function (lines 5 and 7)
- Calculating the slope of the line through the two movable points (line 9)
- Plotting the line through the two movable points (line 11)

Exercise 2.9: Go to Desmos (http://www.desmos.com) and re-create what you see in figure 2.11. You do not need to copy the notes in lines 3, 6, 8, and 10.

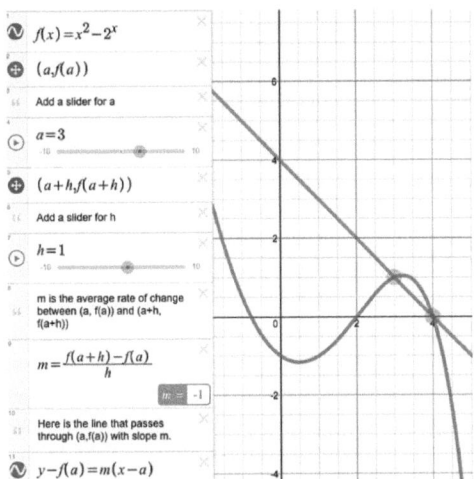

FIGURE 2.11. Using Desmos to find the derivative at a point.

Suppose the position of an object is given by the function $f(x) = x^2 - 2^x$. From the graph in figure 2.11, we can see that the average velocity on the interval $3 \leq x \leq 4$ is -1. We also note that the function appears to be increasing at $x = 3$. That would mean that the instantaneous velocity at $x = 3$ is positive.

Exercise 2.10: Use the slider for h to find an approximate value for the instantaneous velocity at $x = 3$, correct to the tenths place. Record your answers in tables like those shown next.

Approaching 3 from the right

h	$\dfrac{f(3+h)-f(3)}{h}$
1	
0.1	
0.01	
0.001	
0.0001	

Approaching 3 from the left

h	$\dfrac{f(3+h)-f(3)}{h}$
-1	
-0.1	
-0.01	
-0.001	
-0.0001	

What is the slope of the curve, or the derivative of $f(x) = x^2 - 2^x$ at $x = 3$? Answer this question by completing the following statement:
I found that $f'(3) =$ _____. My approximation is correct to _____ decimal places.

If we know a formula for the function, we can use the definition of the derivative to find the derivative of the function at a particular point. Consider the quadratic function

$f(x) = x^2$, which has a graph of a parabola shown in figure 2.10. We may use the definition of the derivative to find $f'(1)$ algebraically.

$$f'(1) = \lim_{h \to 0} \left(\frac{f(1+h) - f(1)}{h} \right)$$

$$= \lim_{h \to 0} \left(\frac{(1+h)^2 - (1)^2}{h} \right) \quad \text{because } f(x) = x^2$$

$$= \lim_{h \to 0} \left(\frac{(1+h)(1+h) - (1)(1)}{h} \right)$$

$$= \lim_{h \to 0} \left(\frac{(1+h+h+h^2) - 1}{h} \right)$$

$$= \lim_{h \to 0} \left(\frac{2h + h^2}{h} \right) \quad \text{by combining like terms in the numerator}$$

$$= \lim_{h \to 0} \left(\frac{h(2+h)}{h} \right) \quad \text{by factoring out } h \text{ in the numerator}$$

$$= \lim_{h \to 0} (2+h) \quad \text{by dividing out } h \text{ in the numerator and denominator}$$

$$= 2 + 0$$

$$= 2$$

Box 2.4: Possible Pitfalls

Recall that if $f(x) = x^2$, then $f(\text{anything}) = \text{anything}^2$.

When carrying out this process, I ask myself at each step if it is permissible to replace h with 0. The answer is no as long as the denominator is h. However, as soon as that h is removed from the denominator, there is no problem.[3] The limit of $2 + h$ as h approaches 0 is the same as $2 + 0$, or 2. Therefore, $f'(1) = 2$.

Consider using this process for the circular ink stain example from the beginning of this section. Use the formula $f(x) = \pi x^2$ and the definition of the derivative to find $f'(0.5)$ algebraically.

$$f'(0.5) = \lim_{h \to 0} \left(\frac{f(0.5+h) - f(0.5)}{h} \right) = \lim_{h \to 0} \left(\frac{\pi(0.5+h)^2 - \pi(0.5)^2}{h} \right)$$

$$= \lim_{h \to 0} \left(\frac{\pi(0.25 + 0.5h + 0.5h + h^2) - 0.25\pi}{h} \right)$$

$$= \lim_{h \to 0} \left(\frac{\pi h + \pi h^2}{h} \right) = \lim_{h \to 0} \left(\frac{h(\pi + \pi h)}{h} \right) = \lim_{h \to 0} (\pi + \pi h) = \pi + \pi(0) = \pi$$

Note that our earlier approximation of a little more than 3 was not that bad! Finally, using the $m = \text{slope } f'(0.5) = \pi$ and the point $(0.5, f(0.5)) = (0.5, 0.25\pi)$, we can use the point-slope formula to write an equation for the tangent line as $y - 0.25\pi = \pi(\pi - 0.5)$.

---

[3] There is no problem as long as I remember that h cannot *be* zero, but it can *approach* zero.

Problem Set 2.2

1. Label points A, B, C, D, E, and F on the graph of $y = g(x)$ in Figure 2.12.

    A is a point on the curve where the derivative is negative.
    B is a point on the curve where the value of the function is negative.
    C is a point on the curve where the derivative is largest.
    D is a point on the curve where the derivative is zero.
    E and F are different points on the curve where the derivative is about the same.

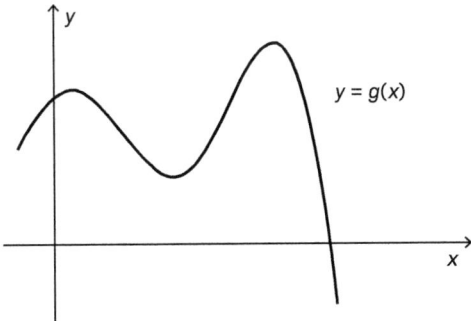

FIGURE 2.12. Graph for problem 1.

2. Use the graph of $y = f(x)$ in Figure 2.13 to fill in the blanks with points A, B, C, D, or E.

    a. $f(4)$ has the same value as the length of the segment with endpoints ___ and ___.

    b. $f(4) - f(1)$ has the same value as the length of the segment with endpoints ___ and ___.

    c. $\dfrac{f(4) - f(1)}{4 - 1}$ has the same value as the slope of the segment with endpoints ___ and ___.

    d. $f'(3)$ has the same value as the slope of the line tangent to $y = f(x)$ at the point ___.

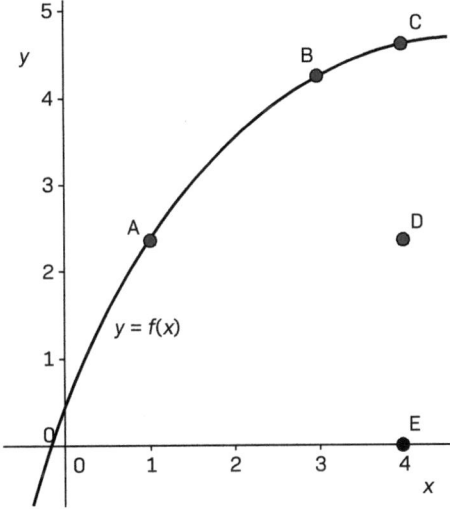

FIGURE 2.13. Graph for problems 2 and 3.

CHAPTER 2 Derivatives | 37

3. Use the graph of $y = f(x)$ in Figure 2.13.
   a. Which is greater: $f(3)$ or $f(4)$? Explain your answer.
   b. Which is greater: $f(3) - f(2)$ or $f(2) - f(1)$? Explain your answer.
   c. Which is greater: $\dfrac{f(2)-f(1)}{2-1}$ or $\dfrac{f(3)-f(1)}{3-1}$? Explain your answer.
   d. Which is greater: $f'(1)$ or $f'(4)$? Explain your answer.

4. The table that follows shows values for $k(x)$ near $x = 4$. Use it to estimate $k'(4)$.

x	3.998	3.999	4.000	4.001	4.002
k(x)	63.904	63.952	64.000	64.048	64.096

5. The table that follows shows values for $g(x)$ near $x = 2$. Use it to estimate $g'(2)$.

x	1.998	1.999	2.000	2.001	2.002
g(x)	1.24845	1.24582	1.24320	1.24059	1.23799

6. The table that follows shows values for $q(x)$ near $x = 3$. Use it to estimate $q'(3)$.

x	2.97	2.98	2.99	3.00	3.01
q(x)	−0.999153	−0.999616	−0.999902	−1.000000	−0.999900

7. Examine the graphs of the functions $f(x) = \dfrac{1}{2}x^2$ and $g(x) = f(x) + 3$ on the same set of axes.
   a. What can you say about the slopes of the tangent lines to the two graphs where $x = 0$? Where $x = 2$? Any point $x = a$?
   b. Explain why adding a constant value, C, to any function does not change the value of the slope of its graph at any point.

8. Refer to the Desmos activity in Figure 2.11. Find the derivative of $f(x) = x^2 - 2^x$ at $x = 3$, correct to 5 decimal places.

9. Refer to the Desmos activity in Figure 2.11. If $f(x) = x^2 - 2^x$, what is the value of $f'(0)$, correct to 2 decimal places?

10. Refer to the Desmos activity in Figure 2.11. For $f(x) = x^2 - 2^x$, at how many places is the derivative of f equal to 0? Estimate those x-values to 1 decimal place.

11. Consider the function $j(z) = 3z^2$.
    a. Find the derivative of $j(z)$ at $z = 10$ algebraically.
    b. Find an equation for the line tangent to $j(z)$ at $z = 10$.

12. Consider the function $p(r) = r^2 + r - 2$.
    a. Find the derivative of $p(r)$ at $r = 1$ algebraically.
    b. Find an equation for the line tangent to $p(r)$ at $r = 1$.

2.3 The Derivative Function

OBJECTIVES FOR SECTION 2.3: Upon completing this section, you will be able to do the following:

- Sketch and interpret the graph of the derivative function of a given function
- Use technology to determine the derivative function of a given function
- Use the definition of the derivative and algebra to find the derivative function of a given function

You may have noticed that, for many functions, the slope of the curve changes. The value of the slope, or the derivative, depends on the *x*-value of the point on the function. In this way, we can think of *the derivative function*. The inputs of the derivative function are the *x*-values that are the domain of the original function. The outputs are the slopes of the curve at that point. Every *x*-value is paired with exactly one slope, so the derivative function is actually a function.

If the original function is $f(x)$, then the derivative function is denoted as $f'(x)$. There is nothing particularly special about the letters f or x. If we were working with $s(t)$, then the derivative would be denoted as $s'(t)$. (In this way, we can say that if $s(t)$ is the position of an object at time t, the velocity of the object at time t is $s'(t)$.)

In the last section, we examined the parabola $f(x) = x^2$ and found the value of the derivative at a few points on the curve. Specifically, we looked at the derivative for three particular values of *x*: −0.5, 0, and 1. Here are the slopes of the curve at those points:

x	−0.5	0	1
f'(x)	−1	0	2

We can use the values in the table to plot three points on a graph, which would be part of the graph of the derivative function $y = f'(x)$.

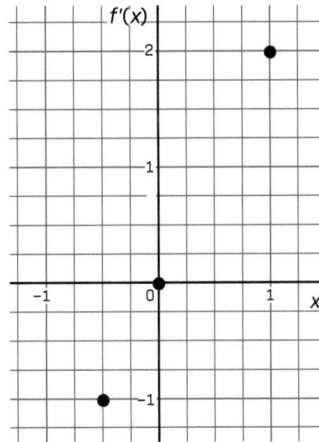

We can use technology to help us determine the derivative of a function at many points and then graph the derivative function.

Exercise 2.11: Here is an activity that uses GeoGebra to graph the derivative of the function $f'(x) = 1 + 3x - x^2$. Perform the numbered steps in GeoGebra (I recommend https://www.geogebra.org/classic), and answer the questions labeled with letters.

1. Open GeoGebra and select the Graphing perspective.
2. Input **f(x)=1+3x-x^2** and press return.
3. Select the slider tool, and then click somewhere in the graphics window, then click OK.
4. Input **A=(a,f(a))** and press return.
5. Click the move tool.
    a. What happens when you move the point on the slider back and forth?
    b. What are the coordinates of point A when $a = 2$?
    c. Is the slope of the curve at point A positive, negative, or zero?
    d. Estimate the value of the slope at point A.
6. Move the slider so that $a = 2$.
7. Input **tangent[A,f(x)]** and press return.
8. Click the slope tool, then click the line from the previous step.
9. Input **B=(a,m)** and press return.
    e. What is the slope of the curve at point A?
    f. What is the value of $f'(2)$?
    g. What are the coordinates of point B?
    h. What happens when you move the point on the slider back and forth?
10. Right-click point B, and then click Show Trace.
    i. What happens when you move the point on the slider back and forth?
    j. What is the value of $f'(0)$?
    k. What is the value of $f'(1)$?
    l. What is the value of x where $f'(x)$ crosses the x-axis? Did you expect this to happen here?
    m. Write an equation for $f'(x)$.
11. Input your equation and press return.
12. Input _____ and press return. This will make GeoGebra graph the derivative function, $f'(x)$. (Your instructor will supply the information in the blank when the time is right.)

Exercise 2.12: Use GeoGebra (and the command from the end of the previous exercise) to graph the derivatives of these functions:

$$f(x) = 2x \qquad f(x) = 2x - 3 \qquad f(x) = 3.5$$

What do you notice about the derivative functions in these cases?

Box 2.5: Teaching Tips

Help your students develop proficiency and understanding by giving them opportunities to practice new techniques with familiar examples. In exercise 2.12, you already know the slopes of the functions, but I am asking you to verify your knowledge using a different method.

Given a graph of a function, it is possible to sketch a graph of the (approximate) derivative function. One strategy is to identify when the derivative is zero, when it is positive, and when it is negative. To improve your sketch, you could estimate the numerical values of the derivative for particular points.

Exercise 2.13: Consider the function $y = f(x)$ shown. We can sketch $y = f'(x)$.

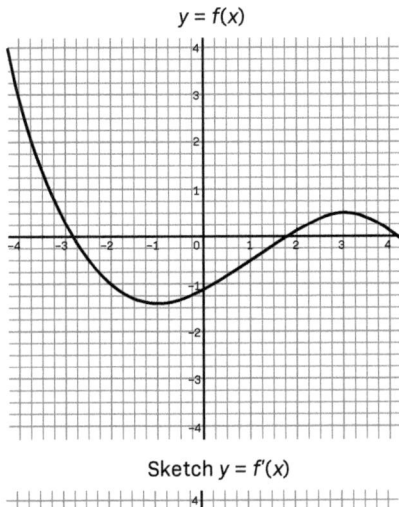

It appears that the slope of the graph $y = f(x)$ is equal to zero for two x-values: -1 and 3. Therefore, we plot points on the derivative graph at $(-1,0)$ and $(3,0)$.

The function f is decreasing for $x < -1$, so the derivative will be negative for $x < -1$.

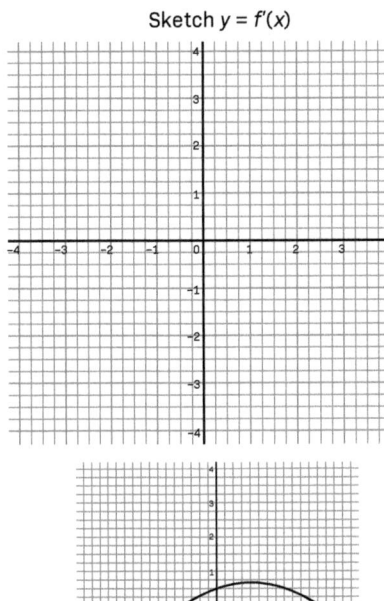

The function f is increasing on $-1 < x < 3$, so the derivative will be positive on the interval $-1 < x < 3$. In fact, it looks like the slope of the curve at $x = 1$ is just a bit less than 1.

Finally, the function f is decreasing for $x > 3$, so the derivative will be negative for $x > 3$.

Our function is continuous and there are no sharp corners, so the derivative will also be continuous. We can anticipate that the derivative will look something like the graph in figure 2.14.

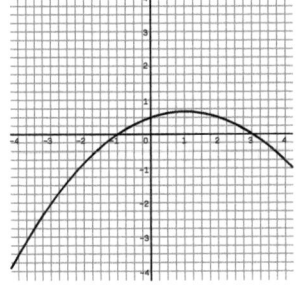

FIGURE 2.14. Graph of the derivative function of $y = f(x)$ in exercise 2.13.

Exercise 2.14: Now try your hand at sketching the graph of the derivative functions of *g* and *h* in figure 2.15.

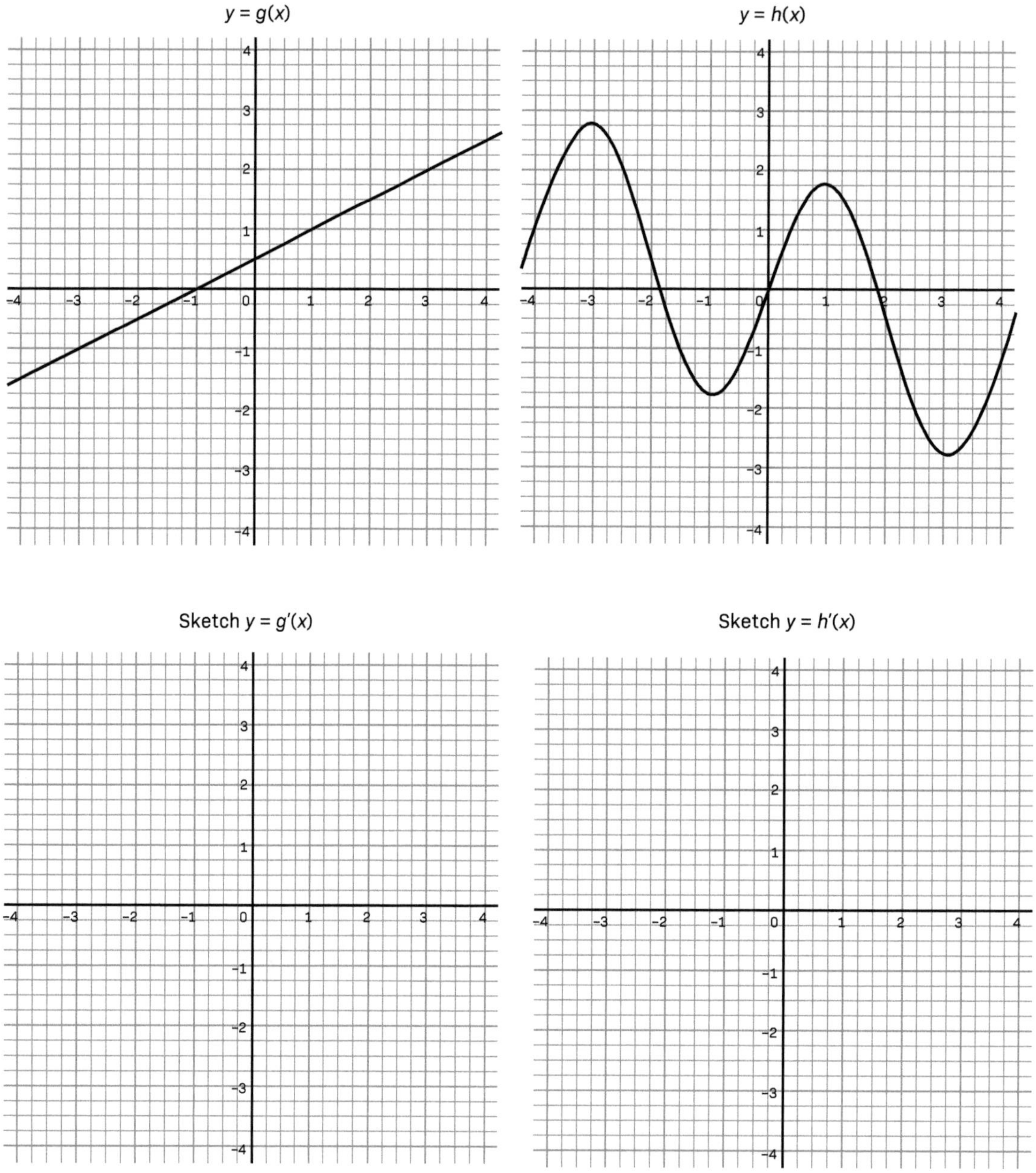

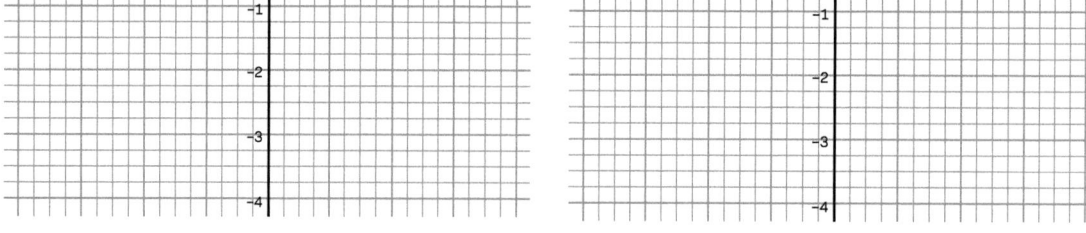

FIGURE 2.15. Graphs for exercise 2.14.

42 | Fundamentals of Calculus for Teachers

> **Box 2.6: Possible Pitfalls**
>
> The vertical axes of the lower graphs represent slopes of the original function, shown in the upper graph. Remember when a function is increasing, its slope is positive at that point. When the function is decreasing, its slope is negative.

The definition of the derivative function is very similar to the definition of the derivative at a point. The main difference is that we consider many values of x instead of one particular value, such as $x = a$.

> **Definition 2.3.1: The Derivative Function**
>
> The derivative of a function $f(x)$ is defined as $f'(x) = \lim\limits_{h \to 0} \left(\dfrac{f(x+h) - f(x)}{h} \right)$
>
> (For now, we will assume this limit exists.)

Let's use this definition to find the formula for the derivative function of $f(x) = x^2$.

$$\begin{aligned}
f'(x) &= \lim_{h \to 0} \left(\frac{f(x+h) - f(x)}{h} \right) \\
&= \lim_{h \to 0} \left(\frac{(x+h)^2 - x^2}{h} \right) \quad \text{because } f(x) = x^2 \\
&= \lim_{h \to 0} \left(\frac{(x+h)(x+h) - x^2}{h} \right) \\
&= \lim_{h \to 0} \left(\frac{(x^2 + xh + xh + h^2) - x^2}{h} \right) \\
&= \lim_{h \to 0} \left(\frac{2xh + h^2}{h} \right) \quad \text{by combining like terms in the denominator} \\
&= \lim_{h \to 0} \left(\frac{h(2x + h)}{h} \right) \quad \text{by factoring out } h \text{ in the numerator} \\
&= \lim_{h \to 0} (2x + h) \quad \text{by dividing out the } h \text{ in the numerator and denominator} \\
&= 2x + 0 \\
&= 2x
\end{aligned}$$

From this, we see that the derivative function is linear, with a slope of 2 and a y-intercept of 0. This is supported by the table of values and graph that we saw at the beginning of this section.

Exercise 2.15: Use the definition of the derivative function to find formulas for the derivatives of the following functions:

a. $s(t) = 2t + 4$
b. $Q(n) = -3$
c. $A(r) = \pi r^2$

Problem Set 2.3

1. Sketch a graph of the derivative function for each function shown.

TABLE 2.5 Graphs for Problem 1

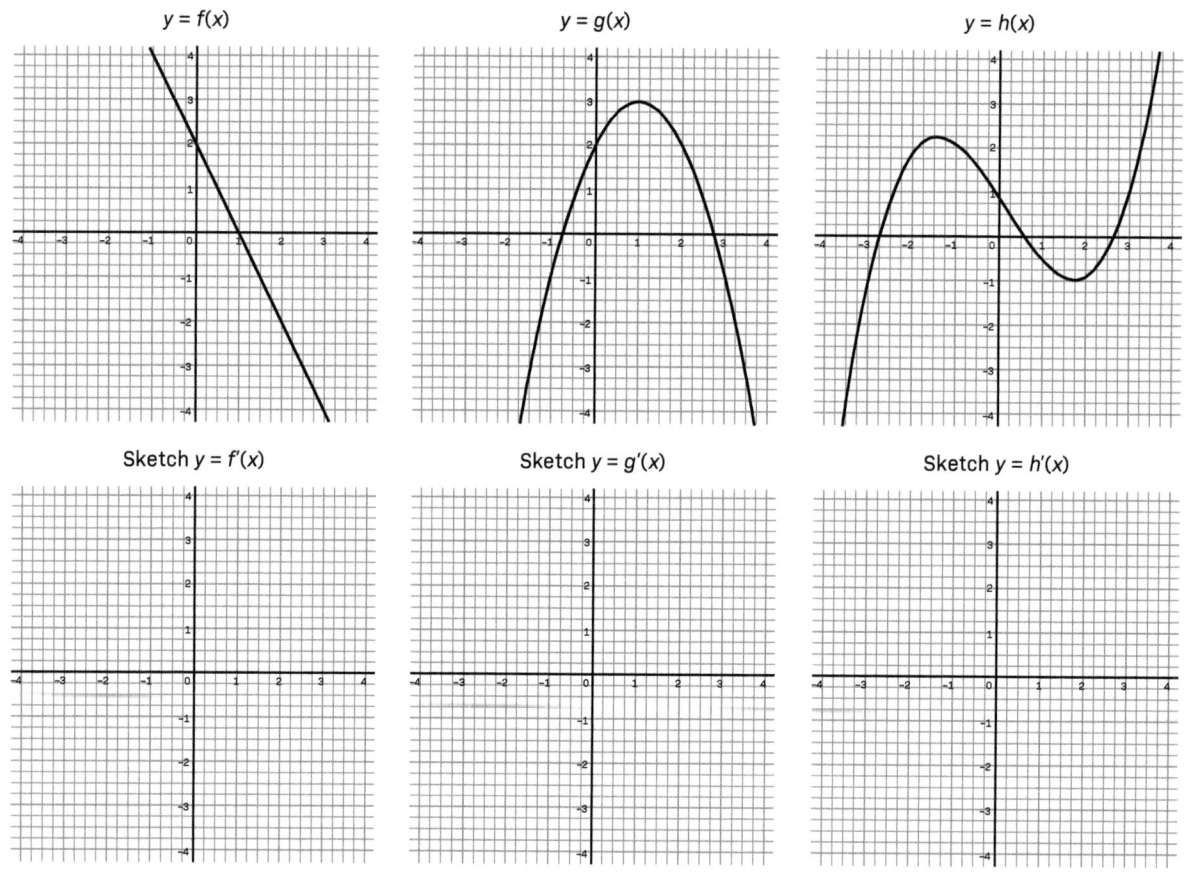

2. Draw a graph of a continuous function $y = f(x)$ that satisfies these three conditions:
   - $f'(x) > 0$ for $x < -2$
   - $f'(x) < 0$ for $-2 < x < 2$
   - $f'(x) = 0$ for $x > 2$

3. In the graph of *f* shown in figure 2.16,
   a. at which of the labeled *x*-values is $f(x)$ greatest?
   b. at which of the labeled *x*-values is $f(x)$ least?
   c. at which of the labeled *x*-values is $f'(x)$ greatest?
   d. at which of the labeled *x*-values is $f'(x)$ least?

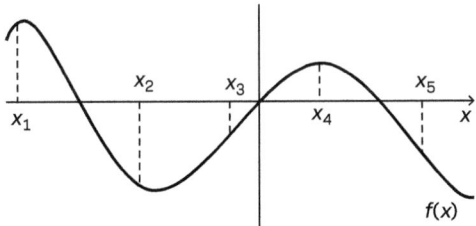

FIGURE 2.16. Graph for problem 3.

4. A child inflates a balloon, admires it for a while, and then lets the air out at a constant rate. If $V(t)$ gives the volume of the balloon at time t, then the graph in figure 2.17 shows $V'(t)$ as a function of t.
   a. At what time does the child begin to inflate the balloon?
   b. At what time does the child finish inflating the balloon?
   c. At what time does the child begin to let the air out?
   d. What would the graph of $V'(t)$ look like if the child had alternated between pinching and releasing the open end of the balloon, instead of letting the air out at a constant rate?

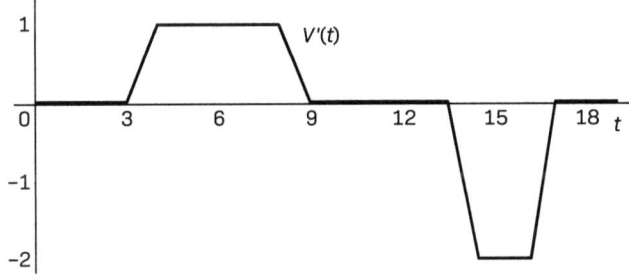

FIGURE 2.17. The rate of change of the volume of a balloon, with respect to time.

5. If $q(r) = 5 - 6r$, find a formula for the derivative function $q'(r)$.
6. If $k(x) = 144$, find a formula for the derivative function $k'(x)$.
7. If $m(y) = 3y^2 - 2$, find a formula for the derivative function $m'(y)$.

2.4 Interpretations of the Derivative

OBJECTIVES FOR SECTION 2.4: Upon completing this section, you will be able to do the following:

- Determine the meaning of the derivative of a function in a given context
- Determine the units of the derivative of a function
- Estimate the value of a function near a given point
- Write and interpret statements and formulas about the derivative of a function

Thus far, we have used the notation $f'(x)$ to denote the derivative of the function $f(x)$. We may interpret $f'(x)$ as the slope of the curve $y = f(x)$, or the slope of the line tangent to $y = f(x)$.

Another notation that is commonly used for the derivative is $\dfrac{dy}{dx}$. This notation is tied closely to the idea of slope. The d is related to the Greek letter delta, Δ. You may recall that Δ is used in science to refer to change. Therefore, we can think of dy as meaning "a small change in the value of y" and dx as "a small change in the value of x." The ratio is interpreted as "a small change in y divided by a small change in x." On a graph of $y = f(x)$, this $\dfrac{dy}{dx}$ refers to the slope of the curve.

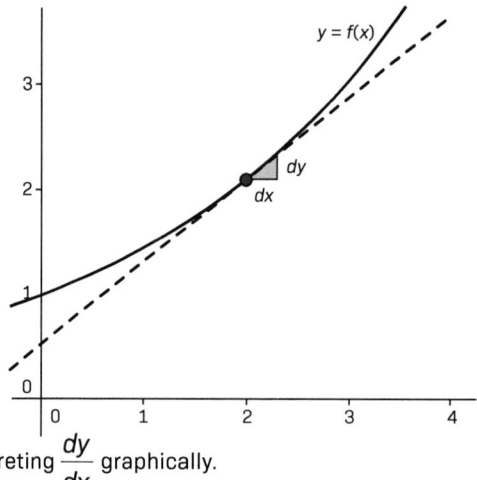

FIGURE 2.18. Interpreting $\dfrac{dy}{dx}$ graphically.

In Figure 2.18, we can denote the derivative of $y = f(x)$ at $x = 2$ as $f'(2)$ or as $\left.\dfrac{dy}{dx}\right|_{x=2}$. As you may notice, one of these is less complicated the other. So, what is the point of using the notation $\dfrac{dy}{dx}$ at all? One reason is the $\dfrac{dy}{dx}$ notation helps us determine the units of the derivative. A second reason is that some technologies, such as TI devices and Desmos, use this notation.

Let's investigate the problem of determining the units of the derivative of a function. We have seen that if x is the input of the function and y is the output, then $\frac{dy}{dx}$ represents the slope of the function. This notation is helpful as we consider the units of the derivative in cases where we have variables other than x and y.

For example, if $s = f(t)$ is the height of an object, measured in feet, at time t, measured in seconds, then the derivative is $f'(t) = \frac{ds}{dt} = \frac{\text{small change in height}}{\text{small change in time}}$. The units for the derivative will be the units of the numerator (feet) divided by the units of the denominator (seconds), which results in "feet per second" and can be denoted as $\frac{\text{ft}}{\text{sec}}$. To interpret the statement $f'(4) = -20$, we need to consider the units of the numbers 4 and -20. The 4 is the input, and will be interpreted as "4 seconds." The -20 is a derivative, with "feet per second" as the units. Therefore, the following three statements are valid interpretations of $f'(4) = -20$:

- At $t = 4$ seconds, the velocity of the object is -20 feet per second.
- At $t = 4$ seconds, the object is moving down at 20 feet per second.
- At $t = 4$ seconds, the object will move down approximately 20 feet over the next second.

Exercise 2.16: Suppose that the function $C = h(m)$ gives the number of calories in a sandwich that contains m grams of mustard.

a. Let's determine the units of $h'(m)$.

We can say that $h'(m) = \frac{dC}{dm}$, so the units will be "_____ per _____", or _____.

b. Which of these are appropriate interpretations of the statement $h'(4) = \frac{1}{2}$?

   i. There is one-half of a calorie in 4 grams of mustard.
   ii. In one-half of a gram of mustard, there are 4 calories.
   iii. When there are 4 grams of mustard, adding another gram of mustard would increase the calories in the sandwich by one-half of a calorie.
   iv. When there are 4 grams of mustard, we would expect approximately 1 calorie more by adding 2 more grams of mustard.

c. Suppose you knew that $h(4) = 550$ and $h'(4) = \frac{1}{2}$. Use this information to estimate $h(6)$.

In the circular ink stain example from figure 2.9, we examined $A = f(r)$, where A was the area of a circle measured in square centimeters and r was the length of the radius measured in centimeters. The units of $f'(r)$ will be the same as the units of

$$\frac{dA}{dr} = \frac{\text{small change in area}}{\text{small change in radius length}},$$

which results in "square centimeters per centimeter" or $\frac{cm^2}{cm}$. At this point, let's resist the urge to "reduce" this fraction. Here's why. Suppose we found that $f'(3) = 9.42$, correct to

2 decimal places. How should we interpret this? The 3 is the input of the function, which is the radius of the circle. Thus, we are considering a circle with a radius of 3 cm. The 9.42 is a derivative, with units "square centimeters per centimeter." Here are two valid interpretations of $f'(3) = 9.42$:

- When the radius of a circle is 3 cm, the area is increasing at a rate of 9.42 cm² per 1 cm increase in the radius length.
- When the radius of a circle is 3 cm, the area will increase by approximately 9.42 square cm if the radius is increased 1 cm.

On Desmos, we can graph the derivative of a function by first typing the equation of the function, and then typing $\frac{d}{dx} f(x)$ in the next line. There is also a button for the derivative in the "misc" tab available from the set of functions, as shown in figure 2.19. Most easily, you may type $f'(x)$.

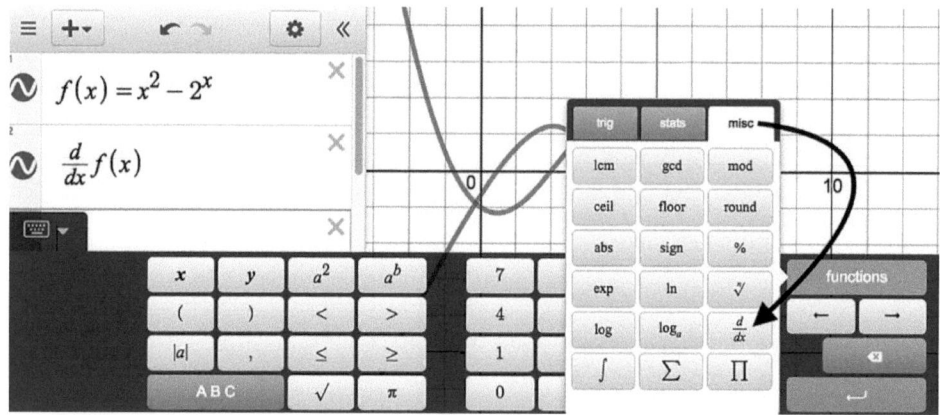

FIGURE 2.19. Using Desmos to graph the derivative function.

Box 2.7: Teaching Tips

Technology changes so quickly that this image may not accurately reflect the current version of Desmos. Stay up to date with advances in educational technology through social media groups like MTBoS.

On a TI-84-Plus calculator, you can press 2ND CALC and choose 6: dy/dx to find the value of the derivative of a function at a point. If you wish to graph the derivative of the function in Y_1, you can use the nDeriv command, found by pressing the MATH key. Depending on your operating system, this will allow you to write nDeriv(Y_1,X,X) or $\frac{d}{dX}(Y_1)\big|_{X=X}$ for Y_2.

With GeoGebra, if the function is listed as f(x), then you can graph the derivative by typing the command **derivative [f]**, or even more simply, **f'(x)**.

Exercise 2.17: Suppose a pie is removed from a hot oven and left on a counter to cool. The function $H = g(t)$ gives the temperature (in degrees centigrade, °C) of the pie after t minutes.

Match the following statements with these symbolic notations:

i. $g(30) = 62$ ii. $g^{-1}(30) = 62$ iii. $g'(30) = -1.4$ iv. $(g^{-1})'(30) = -1.4$

a. The pie reaches 30°C after 62 minutes.
b. When the pie is 30°C, it was about 1°C warmer 1.4 minutes prior to that time.
c. After 30 minutes, the pie is 62°C.
d. After 30 minutes, the pie is cooling about 1.4°C per minute.

Problem Set 2.4

1. An ice cream company knows that the cost, C (in dollars), to produce i quarts of cookie dough ice cream is a function of i, so $C = f(i)$. Suppose you know that $f(200) = 70$ and $f'(200) = 3$.
    a. In $f(200) = 70$ what are the units of the 200? What are the units of the 70? Explain clearly what this equation is telling you.
    b. In $f'(200) = 3$ what are the units of the 200? What are the units of the 3? Explain clearly what this equation is telling you.
    c. Estimate the values of $f(201)$ and $f(198)$.

2. After investing $1,000 at an annual interest rate of 7%, compounded continuously for t years, your balance is B, where $B = f(t)$.
    a. What are the units of $\frac{dB}{dt}$?
    b. Is $\frac{dB}{dt}$ positive, negative, or zero?
    c. What is the financial interpretation of $\frac{dB}{dt}$?

3. Let $f(t)$ be the number of centimeters of rainfall that has fallen since midnight, where t is time in hours.
    a. Write sentences to interpret the following symbols:
        i. $f(10) = 3.1$
        ii. $f'(8) = 0.4$
    b. Write symbolic statements that convey the meaning of the following sentences:
        i. A total of 2 cm of rain had fallen between midnight and 7 a.m.
        ii. At 11 a.m., the rain was falling at a rate of 0.6 cm per hour.

2.5 The Second Derivative

OBJECTIVES FOR SECTION 2.5: Upon completing this section, you will be able to do the following:

- Interpret tables and graphs with respect to the second derivative of a function
- Relate the concavity of the graph of a function to the second derivative
- Use technology and algebra to determine formulas for the first and second derivatives of a function.

How is the slope of a function changing? What is the rate of change of the rate of change? What is the derivative of the derivative? This is known as the *second derivative* of a function. It may be denoted as $f''(x)$ or $\dfrac{d^2y}{dx^2}$.

> **Box 2.8: Possible Pitfalls**
>
> If you feel that you are just beginning to understand the first derivative and are worried about examining the second derivative, you are not alone. Take a deep breath and relax. We're in this together. This is a good time to chat with a classmate about derivatives. Share what you do know now, and what you *want to know*.

Recall the context from problem set 2.3 in which a child was inflating a balloon. In that context, $V(t)$ gave the volume of the balloon at time t. You were asked to interpret the graph of $V'(t)$, the derivative of the volume function, shown in Figure 2.20.

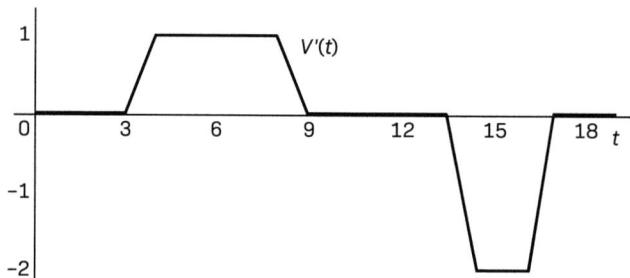

FIGURE 2.20. The rate of change of the volume of a balloon, with respect to time.

We could use this graph to examine how the derivative is changing. There are several intervals where the graph is a horizontal line. For example, at $t = 6$, the slope of $V'(t)$ is zero, so we may say $V''(6) = 0$. Similarly, $V''(10) = 0$ and $V''(15) = 0$. At other points, such as where $t = 8.5$, the graph of $V'(t)$ is decreasing. Therefore, we could say that $V''(8.5)$ is negative. By similar reasoning, $V''(3.5)$ would be positive.

We can see an application of the second derivative in the concept of acceleration, because acceleration is the rate of change of velocity, and velocity is the rate of change of position.

Symbolically, if s(t) gives the position of an object at time t, then the velocity of the object at time t is v(t) = s′(t). Acceleration is the first derivative of velocity, and therefore the second derivative of the position, so we may represent the acceleration at time t as a(t) = v′(t) = s″(t).

Examine the graphs of distance vs. time in Figure 2.21. Note that all three are increasing, so the first derivative, or velocity, is positive in all three graphs. How is the velocity changing in these three examples? This question speaks to the acceleration, which is the second derivative of the distance function with respect to time.

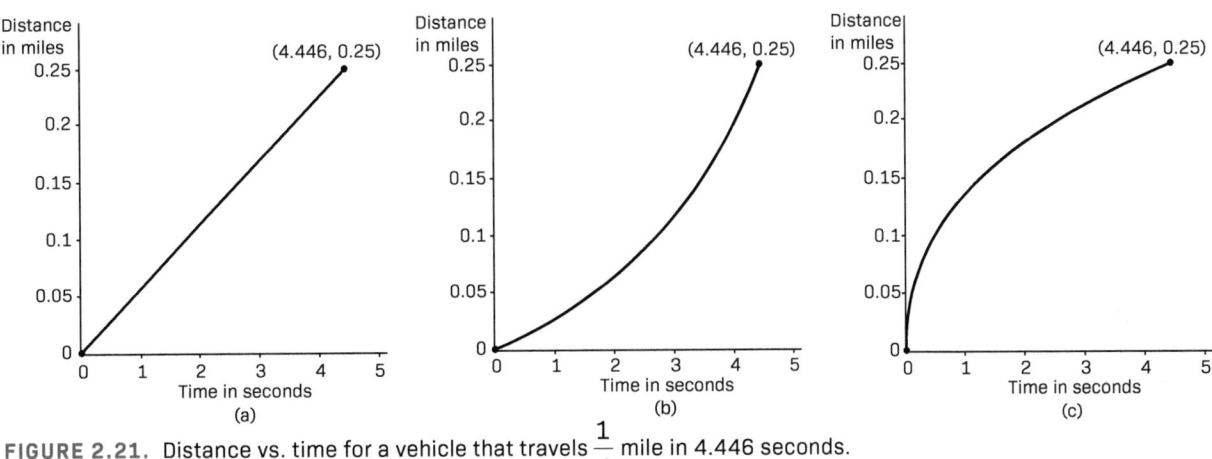

FIGURE 2.21. Distance vs. time for a vehicle that travels $\frac{1}{4}$ mile in 4.446 seconds.

In the graph on the left (figure 2.21.a), the straight line indicates the velocity is constant throughout the entire 4.446-second interval. The velocity does not change, and therefore there is no acceleration. In other words, the second derivative of this linear function is zero.

In the graph on the middle (figure 2.21.b), the velocity is increasing over the interval. Therefore, the derivative of the velocity, the acceleration, is positive. We can say that the second derivative of this function shown in this graph is positive.

Finally, in the graph on the right, the velocity is decreasing over the interval, although it remains positive. In this case, the acceleration is negative, and the second derivative of the function shown is negative.

Exercise 2.18: Draw distance vs. time graphs that show the following:

a. Negative velocity and positive acceleration
b. Negative velocity and zero acceleration
c. Negative velocity and negative acceleration

Exercise 2.19: Use Desmos to graph the function $f(x) = x^3 - 3x$.

a. List the value(s) of x for which the derivative of the function is zero.
b. List the interval(s) of x where the derivative of the function is positive.
c. List the interval(s) of x where the derivative is increasing.
d. List the interval(s) of x where the derivative is decreasing.
e. List the interval(s) of x where the function is concave up.
f. List the interval(s) of x where the function is concave down.

Exercise 2.20: Suppose f is a function that has a first and second derivative. Summarize your findings about the first and second derivative by completing the following statements with words *positive* or *negative*.

a. If f is increasing on an interval, then f' is _____ on that interval.
b. If f is decreasing on an interval, then f' is _____ on that interval.
c. If f' is increasing on an interval, then f'' is _____ on that interval.
d. If f' is decreasing on an interval, then f'' is _____ on that interval.
e. If f is concave up on an interval, then f'' is _____ on that interval.
f. If f is concave down on an interval, then f'' is _____ on that interval.

A function has an *inflection point* where the concavity changes from concave up to concave down. This occurs precisely where the derivative is the greatest (or least) along that interval. In the function $f(x) = x^3 - 3x$, there is an inflection point at $x = 0$, where the function changes from concave down to concave up.

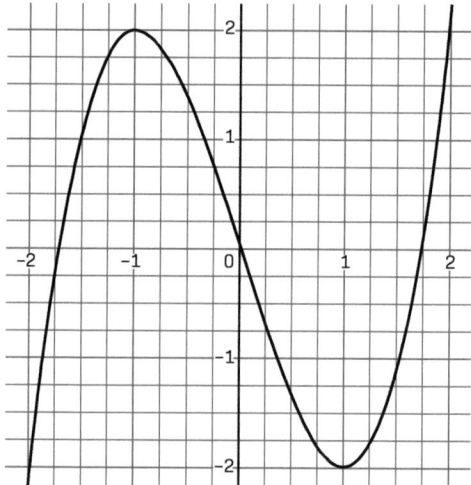

FIGURE 2.22. A function with an inflection point.

Problem Set 2.5

1. The graph the function $f(x)$ is shown in figure 2.23. Complete the table to indicate the sign of the function, the first derivative, and the second derivative at each point. That is, tell whether f, f', and f'' are positive, negative, or zero.

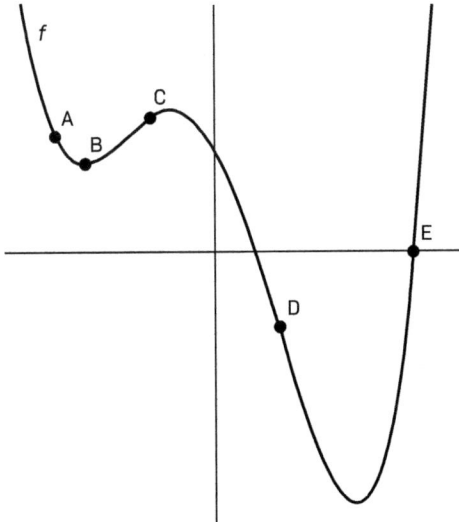

POINT	f	f'	f''
A	+	−	+
B			
C			
D			
E			

FIGURE 2.23. Graph and table for problem 1.

2. The graph of g' (not g) is given in figure 2.24. At which of the marked values of x is
    a. $g'(x)$ the greatest?
    b. $g'(x)$ the least?
    c. $g''(x)$ the greatest?
    d. $g''(x)$ the least?
    e. $g(x)$ the greatest?
    f. $g(x)$ the least?

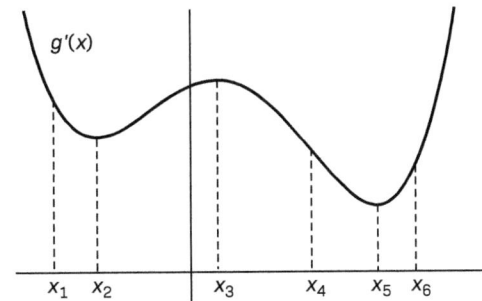

FIGURE 2.24. Graph of g' for problem 2.

3. Which of the points labeled by letters in the graph of *f* in figure 2.25 have
   a. f' and f'' nonzero and of the same sign?
   b. at least two of f, f', and f'' equal to zero?

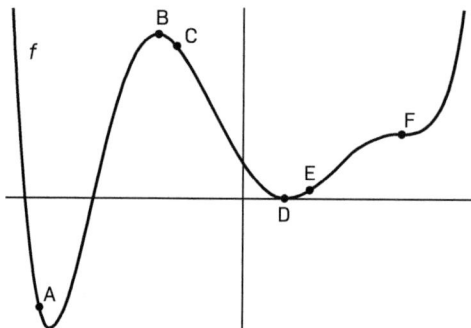

FIGURE 2.25. Graph for problem 3.

4. Scientists drop a probe into the ocean from a ship. The depth of the probe, measured in kilometers, is given by $s(t) = t^2 + 2t$, where *t* is measured in hours after dropping the probe.
   a. Write a formula for the probe's velocity at time *t*.
   b. What are the units of the velocity in this example?
   c. Write a formula for the probe's acceleration at time *t*.
   d. What are the units of the acceleration in this example?

References

Common Core State Standards Initiative. (n.d.). Retrieved from http://www.corestandards.org

LitbyNoonProductions. (2012, August 23). 337 58 MPH TFD DSR Tony Schumacher Alan Johnson Brainerd Mn 2005 [Video file]. Retrived from https://www.youtube.com/embed/rYNeGbOwvnY?start=120&end=155

Meyer, D. (2000). Basketball strobes [Video file]. Retrieved from https://vimeo.com/16832687

Texas Education Agency. (n.d.). Texas education knowledge and skills. Retrieved from https://tea.texas.gov/curriculum/teks/

Weierich, T. (2013, September 18). Vertical jump in slow motion [Video file]. Retrieved from https://www.youtube.com/embed/rYNeGbOwvnY?start=120&end=155

Credits

IMG 2.2: Copyright © by Desmos, Inc.
Fig. 2.11: Copyright © by Desmos, Inc.
Fig. 2.19: Copyright © by Desmos, Inc.

Integrals

Concepts

Think

If you knew the speed of a vehicle and the time that it was traveling at that speed, could you find the total distance that it traveled? The formula *distance = rate × time* would be very useful. Now think about the real-life situations where the speed isn't constant and the vehicle speeds up, slows down, and occasionally stops. How would you use *distance = rate × time* to find the total distance traveled in this situation? In this chapter, we will begin with this situation and see how calculus helps us approximate the total distance traveled as an accumulation of distances traveled over short periods of time. This accumulation involves putting things together, and the mathematical term for this is *integration*. Later in the chapter, we will show how the concept of integration can help us find areas of shapes, including shapes that don't have straight edges. The ideas of accumulation and integration extend our discussion of approximations and limits from the first chapter.

Remember

To prepare for this chapter, you may want to think about the following topics. They serve as prerequisites and primers to the mathematical content in the chapter.

- Determine the distance that a vehicle travels, given the speed over a period of time—for example, a car that travels at a constant speed of 75 miles per hour for 20 minutes.
- Calculate the area of a rectangle, given the measurements of base and height, such as a rectangle that is 2 units tall and 0.25 units wide, or a rectangle that is k units tall and w units wide.
- Describe how a geometric figure can be formed by putting together, or taking apart, familiar geometric figures, such as combining a rectangle and semicircle to make the letter P.

Connect

In the middle grades, the ideas of accumulation, composition, and decomposition appear within the context of geometry and measurement. Specifically, the Texas Essential Knowledge and Skills (Texas Education Agency, n.d.) states that students in grade 6 will decompose and rearrange parallelograms, trapezoids, and triangles to make sense of area formulas. In grade 7, students use these formulas to find areas of composite figures made from various familiar shapes. The Common Core State Standards for Mathematics (CCSS-M) (n.d.) have similar standards at these grade levels.

3.1 How Do We Measure Distance Traveled?

OBJECTIVES FOR SECTION 3.1: Upon completing this section, you will be able to do the following:

- Approximate the distance traveled over an interval
- Calculate left- and right-hand sums
- Interpret and create graphs of Riemann sums
- Make approximations to a specified level of accuracy

We began our discussion in the previous chapter with a question about how to measure speed when we are given data on the position of an object at various times. In this chapter, we will begin with a related but different question: How do we measure the distance traveled when we have data on the velocity of an object at various times?

If the velocity of the object is constant, then this is fairly straightforward application of $d = rt$, where distance is the product of rate and time. For example, if a car was driving along a highway at 60 miles per hour for 2 hours, the car would have traveled 120 miles during that time. Similarly, a car that drives 30 miles per hour for four minutes travels 2 miles, as we see, here:

$$\left(30\frac{\text{miles}}{\text{hour}}\right) \cdot \left(\frac{4}{60}\text{hour}\right) = 2 \text{ miles}$$

But what if the velocity isn't constant? In city traffic, and in regular driving, it is rare that the speed stays constant. Braking and accelerating make the speed of the car change over time. In this section, we will investigate how to estimate the distance traveled given the velocity of a vehicle.

Suppose that a passenger in your car was keeping track of the speed, and recording it every four minutes. (Actually, the computer in your car may do this for you.) The data recorded states that at the beginning your speed was 24 miles per hour, at four minutes your speed was 30 miles per hour, and at eight minutes your speed was 60 miles per hour. For now, let's also suppose that your speed increased over this time period. How far did you travel during these eight minutes?

While we can't determine this value exactly, we can make some estimates. One way to estimate would be to assume that the velocity was constant for the four-minute intervals.

- Here's one estimate. If the velocity was 24 miles per hour for the first four minutes, and 30 miles per hour for the next four minutes, the distance traveled would be

$$\left(24\frac{\text{miles}}{\text{hour}}\right)\cdot\left(\frac{4}{60}\text{hour}\right)+\left(30\frac{\text{miles}}{\text{hour}}\right)\cdot\left(\frac{4}{60}\text{hour}\right)=1.6\text{ miles}+2\text{ miles}=3.6\text{ miles}$$

- Here's another estimate. If the velocity was 30 miles per hour for the first four minutes, and 60 miles per hour for the next four minutes, the distance traveled would be

$$\left(30\frac{\text{miles}}{\text{hour}}\right)\cdot\left(\frac{4}{60}\text{hour}\right)+\left(60\frac{\text{miles}}{\text{hour}}\right)\cdot\left(\frac{4}{60}\text{hour}\right)=2\text{ miles}+4\text{ miles}=6\text{ miles}$$

Based on these two estimates, we could say that the actual distance traveled during the eight-minute period is between 3.6 miles and 6 miles. Because your speed increased over the time interval, the first value is an underestimate and the second is an overestimate.

Exercise 3.1: Suppose we measure a car's velocity every two seconds and obtain the data in the following table:

TIME (SEC)	0	2	4	6	8	10
VELOCITY (FT/SEC)	10	28	42	52	58	60

a. How far has the car traveled in the 10-second interval?
   Underestimate:
   Overestimate:
b. Name two ways that we can get a better estimate.

In exercise 3.1, the car's velocity is increasing over the 10-second interval. If we divide the 10-second interval into five 2-second intervals and use the velocity from the left-hand endpoint of each interval, we will underestimate the total distance. Similarly, if we use the velocity from the right-hand endpoint of the interval, we will overestimate the distance traveled during the 10 seconds.

If the velocities were decreasing over the entire interval, then the opposite phenomena would occur. Using the left-hand endpoints would overestimate the distance because they would be larger than those at the right-hand side; using the right-hand endpoints would underestimate the total distance.

Box 3.1: Possible Pitfalls

Use a picture, like figure 3.1, to determine whether the sum is an underestimate or overestimate.

If, instead, the velocities were sometimes increasing and sometimes decreasing, then it would be difficult to tell whether the left-hand side or right-hand side was the underestimate without actually calculating. Fortunately, our purpose is not to determine whether we have underestimated or overestimated the distance. Instead, we wish to compare the two estimates and improve them. One way to do this is by increasing the number of intervals, which decreases the length of time for each interval.

Exercise 3.2: Examine the data of a car's velocity, measured every second, in the following table:

TIME (SEC)	0	1	2	3	4	5	6	7	8	9	10
VELOCITY (FT/SEC)	10	19	28	35	42	47	52	55	58	59	60

Estimate the distance that the car traveled during this 10-second interval

a. using left-hand endpoints, and
b. using right-hand endpoints.

We can draw rectangles to illustrate the estimates that we obtained earlier. For each interval, we found the distance by multiplying the velocity by the time (i.e., the width of the interval); this corresponds to the area of a rectangle. The total distance is the sum of the areas of these rectangles.

In general, if we have a function $v(t)$ that gives the velocity of an object at time t, we can estimate the distance traveled over the interval $a \leq t \leq b$ (that is, from $t = a$ to $t = b$). We could divide the entire interval into n equal sub-intervals. We use Δt to refer to the width of the sub-intervals, where $\Delta t = \dfrac{\text{width of entire interval}}{\text{number of sub-intervals}} = \dfrac{b-a}{n}$. We can name these individual points along the t-axis $t_0, t_1, t_2, t_3, \ldots t_{n-1}, t_n$, where $t_0 = a$ and $t_n = b$.

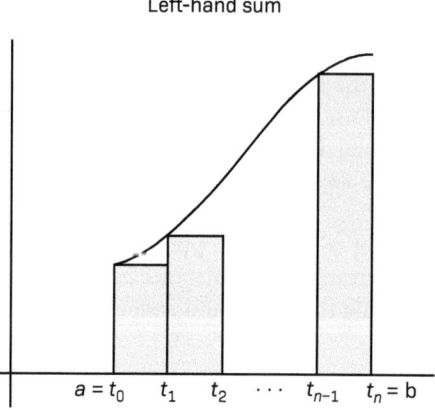 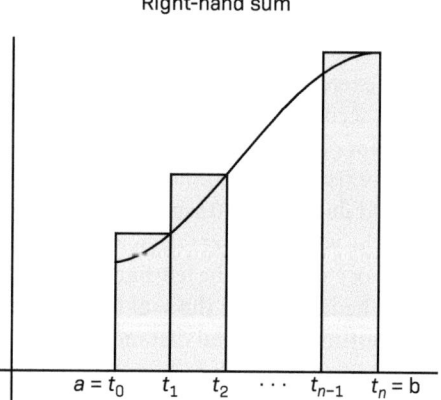

FIGURE 3.1. Left-hand sum and right-hand sum.

Use the picture in figure 3.1 and the notation for the left-hand sum to see if you can make sense of the formula for the right-hand sum.

$$\text{Left-hand sum} = \sum_{i=0}^{n-1}\left(v(t_i)\cdot \Delta t\right) = v(t_0)\cdot \Delta t + v(t_1)\cdot \Delta t + v(t_2)\cdot \Delta t + \cdots + v(t_{n-1})\cdot \Delta t$$

$$\text{Right-hand sum} = \sum_{i=1}^{n}\left(v(t_i)\cdot \Delta t\right) = v(t_1)\cdot \Delta t + v(t_2)\cdot \Delta t + v(t_3)\cdot \Delta t + \cdots + v(t_n)\cdot \Delta t$$

We can apply these ideas to the velocity versus time data to visualize the estimates of the distance traveled. Figure 3.2 shows a graph of the velocities, and the horizontal axis has been marked with two-second intervals, as in the case of exercise 3.1.

Box 3.2: Possible Pitfalls

Remember that $\Delta t = \dfrac{b-a}{n}$.

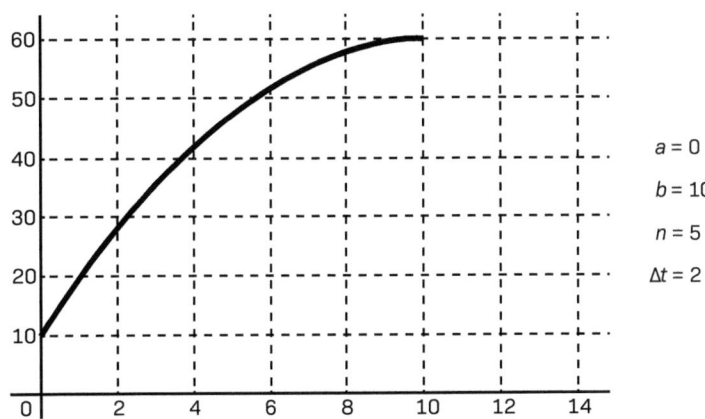

FIGURE 3.2. Using velocity data to approximate distance traveled.

In this particular example, we could draw rectangles that each have a base of two seconds and heights that correspond to velocities at the left endpoint of each interval. The first rectangle would have a base from 0 to 2 seconds, and a height of 10 feet per second. The second rectangle would have a base from 2 to 4 seconds, and a height of 28 feet per second. This continues until we make the fifth rectangle, which has a base from 8 to 10 seconds and a height of 58 feet per second.

Exercise 3.3: Does this picture, using left-hand endpoints, represent an overestimate or underestimate? How can you tell?

Using a different color, we can draw in rectangles that use the right-hand endpoints. That is, the first rectangle would have a base from 0 to 2 seconds and a height of 28 feet per second. The fifth rectangle would have a base from 8 to 10 seconds and a height of 60 feet per second.

Exercise 3.4: Does this picture, using right-hand endpoints, represent an overestimate or underestimate? How can you tell?

Exercise 3.5: Use the graph in figure 3.3 to make a drawing to estimate the distance traveled for the data that were collected every second (see exercise 3.2).

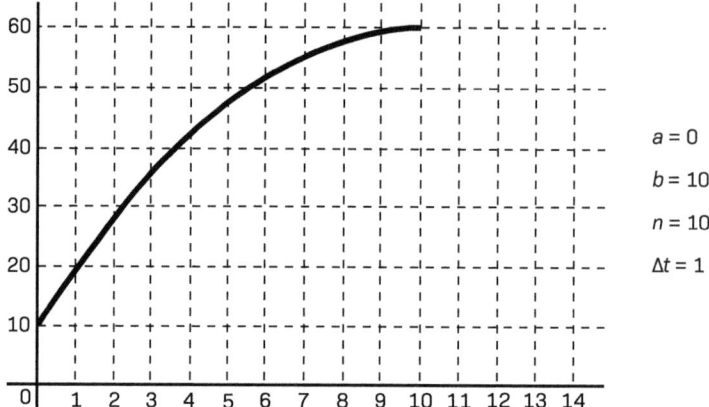

FIGURE 3.3. Make a drawing to estimate the distance traveled.

You should notice that the difference between the upper and lower estimates decreases when the number of intervals increases. We can illustrate this with drawings like figures 3.2 and 3.3. For each time interval, the difference between the two estimates is a small rectangle. We can "stack" all of these differences together to make a single rectangle. The width of this rectangle is the same as the width of the individual rectangles. The height is the difference between the velocity at the end and the velocity at the beginning. In figure 3.4, you will see examples of this for the cases of $\Delta t = 2$ and $\Delta t = 1$. (You can investigate other cases in the GeoGebra applet "3.1 Left- and Right-Hand Sums: Difference" at https//ggbm.at/BN8D2duN.)

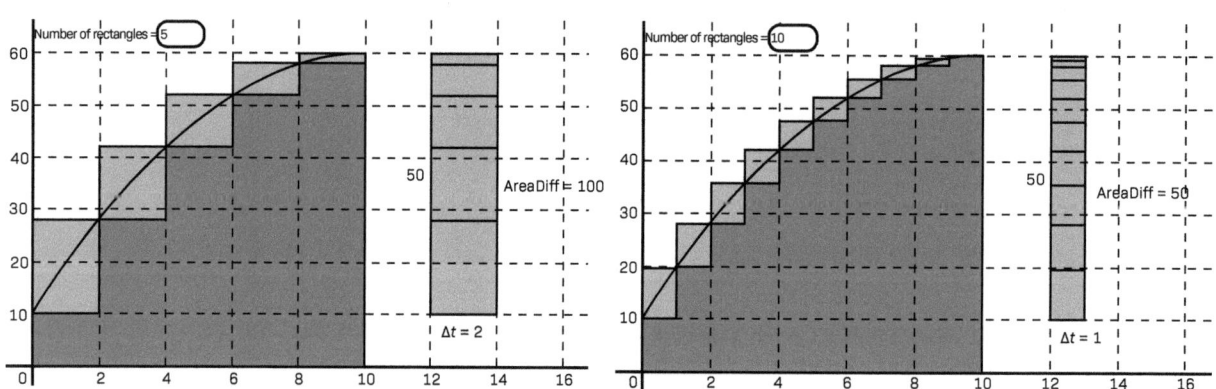

FIGURE 3.4. Differences of left-hand sums and right-hand sums.

If we want to make the difference between estimates small, we must change something. We can't change the velocities at the endpoints, but we can change the width of the rectangles. Using smaller rectangles would decrease the difference between upper and lower estimates.

In a case where there are increasing velocities, the difference between the right-hand and left-hand sum would be

$$\text{Right-hand sum} - \text{Left-hand sum} = \sum_{i=1}^{n}(v(t_i) \cdot \Delta t) - \sum_{i=0}^{n-1}(v(t_i) \cdot \Delta t)$$

$$= (v(t_1) \cdot \Delta t + v(t_2) \cdot \Delta t + v(t_3) \cdot \Delta t + \cdots + v(t_n) \cdot \Delta t)$$
$$\quad - (v(t_0) \cdot \Delta t + v(t_1) \cdot \Delta t + v(t_2) \cdot \Delta t + \cdots + v(t_{n-1}) \cdot \Delta t)$$
$$= v(t_n) \cdot \Delta t - v(t_0) \cdot \Delta t$$
$$= (v(t_n) - v(t_0)) \cdot \Delta t$$
$$= (v(b) - v(a)) \cdot \Delta t$$

Exercise 3.6: Consider the graph presented in figure 3.1.

a. What would be the difference between the upper and lower estimates if the data on the velocity were given
  i. every tenth of a second?
  ii. every hundredth of a second?
  iii. every thousandth of a second?
b. How frequently must the velocity be recorded to estimate the total distance traveled to within 0.1 feet?

With this in mind, we can improve the accuracy of our estimates by decreasing the value of Δt, the width of each time interval. This, in turn, increases n, the number of intervals that we will examine. There is a definite trade-off at work here. The more intervals we have, the more accurate the results, and the more computation is involved. If we know in advance how accurate we would like our estimates, then we can potentially save some work by determining the number of intervals needed to achieve it.

Problem Set 3.1

1. A car comes to a stop six seconds after a driver applies the brakes. While the brakes are applied, the following velocities are recorded:

TIME (SEC)	0	2	4	6
VELOCITY (FT/SEC)	90	60	30	0

a. Give lower and upper estimates for the distance that the car traveled after the brakes were first applied.
b. On a sketch of the graph of velocity against time, show the upper and lower estimates from part a.

2. Roscoe was driving on the highway when he saw a sign that said **ROAD ENDS 350 FEET**. He slammed on his brakes and came to a stop in 8 seconds. Below is a table of data from his vehicle's computer while he was braking.

TIME (SEC)	0	2	4	6	8
VELOCITY (FT/SEC)	92	66	42	20	0

   a. Given the data on the table, what is your best estimate for the total distance Roscoe traveled before coming to a stop?

   b. Which of the following statements is true? Explain your reasoning.

      i. Roscoe came to a stop before the road ended.

      ii. We can't tell. Roscoe may or may not have gone off the end of the road.

      iii. Roscoe came to a stop but went off the end of the road.

3. Calvin likes to run. Use the graph of Calvin's velocity in figure 3.5 to estimate the total distance he traveled over a three-hour period.

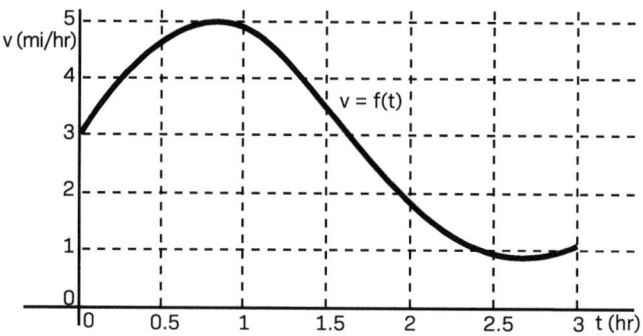

FIGURE 3.5. Calvin's velocity during a three-hour run.

3.2 The Definite Integral

OBJECTIVES FOR SECTION 3.2: Upon completing this section, you will be able to do the following:

- Understand the relationship between left- and right-hand sums and the definite integral of a function over an interval
- Interpret the standard notation of a definite integral
- Use definite integrals and technology to find signed areas

In the last section, we investigated a method to estimate the distance traveled based on velocity data. The method we used involved finding the sum of areas of rectangles. We may apply this to other contexts besides those related to velocity and distance. These sums of rectangular areas approximate the area under a curve. In particular, the left-hand sum and the right-hand sum both approach the actual value with increasing accuracy as the number of rectangles increases.

For a function $f(t)$ that is continuous on the interval $a \leq t \leq b$, we may divide the interval into n equal subintervals, each with a width of Δt, and find the left-hand sum and right-hand sum.

$$\text{Left-hand sum} = \sum_{i=0}^{n-1}\bigl(f(t_i)\cdot \Delta t\bigr) = f(t_0)\cdot \Delta t + f(t_1)\cdot \Delta t + f(t_2)\cdot \Delta t + \cdots + f(t_{n-1})\cdot \Delta t$$

$$\text{Right-hand sum} = \sum_{i=1}^{n}\bigl(f(t_i)\cdot \Delta t\bigr) = f(t_1)\cdot \Delta t + f(t_2)\cdot \Delta t + f(t_3)\cdot \Delta t + \cdots + f(t_n)\cdot \Delta t$$

These types of sums are called *Riemann sums*, named for the 19th-century mathematician G. F. Bernhard Riemann. As the number of subintervals increases, the accuracy of the estimate also increases, and the left-hand sum and the right-hand sum converge[1] to the same value. This value is known as the *definite integral*.

Definition 3.2.1: Definite Integral

The definite integral of f on the interval $a \leq t \leq b$ is written $\int_a^b f(t)\,dt$. At times, this is called the definite integral of f from a to b.

The definite integral of f from a to b is the limit of the left-hand sum (or right-hand sum) with n subdivisions of the interval from a to b as n gets large.

$$\int_a^b f(t)\,dt = \lim_{\text{number of subdivisions} \to \infty} (\text{Left-hand sum}) = \lim_{n \to \infty}\left[\sum_{i=0}^{n-1}\bigl(f(t_i)\cdot \Delta t\bigr)\right] \text{ and}$$

$$\int_a^b f(t)\,dt = \lim_{\text{number of subdivisions} \to \infty} (\text{Right-hand sum}) = \lim_{n \to \infty}\left[\sum_{i=1}^{n}\bigl(f(t_i)\cdot \Delta t\bigr)\right]$$

The function, $f(t)$, is called the *integrand*.

The values a and b are called the *limits of integration*. The lower limit is a and the upper limit is b.

The integral sign $\int$ looks like an elongated S. This reflects the idea of summation, represented in the Riemann sum by Σ.

The dt is called an *infinitesimal* and refers to a small change in t. This reflects the idea of the Δt in the Riemann sum.

---

[1] This description assumes that the definite integral exists.

Together, $\int$ and dt act like grouping symbols, similar to parentheses, to show where the definite integral statement begins and ends.

Exercise 3.7: Use the graphs in figure 3.6 to calculate the left-hand and right-hand sums with $n = 2$ and $n = 10$ to estimate $\int_{1}^{2} \frac{1}{t} dt$. Are these overestimates or underestimates of the exact value of the definite integral?

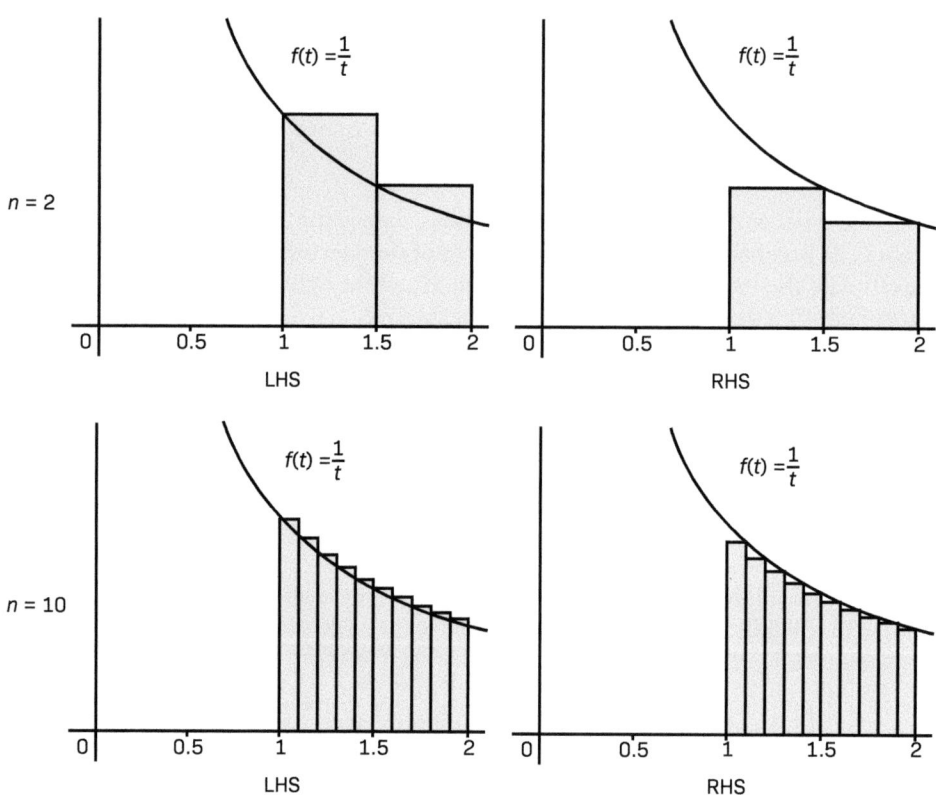

FIGURE 3.6. Estimate the value of a definite integral.

Box 3.3: Teaching Tips

Technology will do a lot for students, but they need to understand the results that technology provides. Let them practice with simple numbers by hand to help them understand the process, then use technology to tackle the complicated situations. To help you understand the results you'll obtain in GeoGebra, use a pencil and calculator to determine these sums.

We can use GeoGebra to draw and calculate the left-hand sums, right-hand sums, and other various Riemann sums. The function RectangleSum has five inputs, as shown:

Rectanglesum[<Function> , <Start x-Value> , <End x-Value> , <Number of Rectangles> , <Position for Rectangle Start>]

When **<Position for Rectangle Start>** = 0, GeoGebra provides the left-hand sum; using a 1 for this value will yield the right-hand sum.

Exercise 3.8: Investigate these commands for $n = 2$ and $n = 10$ with the GeoGebra applet "3.2 Rectangle Sums" at https://ggbm.at/BN8D2duN.

Exercise 3.9: Use left and right sums with $n = 250$ (and technology!) to estimate the value of $\int_1^2 \frac{1}{t}\,dt$. Complete the following statements:
  a. Using 250 rectangles, I estimate the value of $\int_1^2 \frac{1}{t}\,dt$ to be _____.
  b. This approximation is correct to _____ decimal places.

GeoGebra also has the Integral command, which has three inputs. As you may suspect, it will provide the value of the definite integral of a function on an interval. The command is illustrated:

Integral[<Function> , <Start x-Value> , <End x-Value>]

As you may have noticed, the definite integral can be viewed as an area. More specifically, when $f(x)$ is positive on the interval $a \leq x \leq b$ and $a < b$, $\int_a^b f(x)\,dx$ describes the area that is
- between the vertical lines $x = a$ and $x = b$ and
- between the x-axis and the graph of $y = f(x)$.

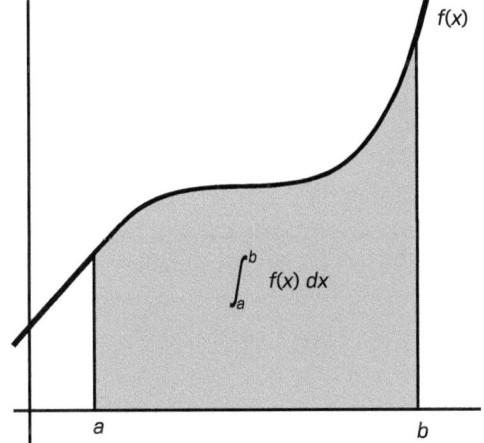

FIGURE 3.7. The definite integral as an area.

Exercise 3.10: Consider the definite integral $\int_{-2}^{2} \sqrt{4-x^2}\, dx$.

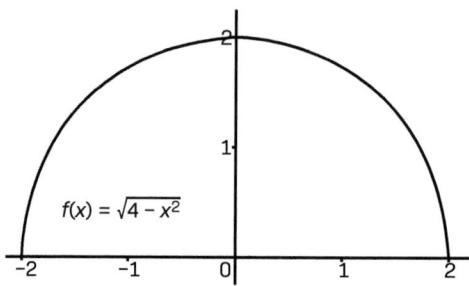

a. Interpret the integral as an area and find its exact value.
b. Estimate the integral using a calculator or computer. Compare your answers to the exact value.

A part of the earlier description of the definite integral as an area included the qualification that the function $f(x)$ be positive on an interval. What about those cases when $f(x)$ is not positive?

Exercise 3.11: To address this question, use technology to see how the definite integral $\int_{-1}^{1} (x^2 - 1)\, dx$ relates to the area between the parabola $y = x^2 - 1$ and the x-axis. What do you find?

You should see that the value of the definite integral in this case is negative and is therefore the opposite of the area of the region between the parabola and the x-axis between $x = -1$ and $x = 1$. This occurs because the "heights" of the rectangles are negative. Therefore, the definite integral can be interpreted as a *signed area*, with positive values indicating areas above the x-axis, and negative values indicating areas below the x-axis.

Box 3.4: Possible Pitfalls

Remember that the definite integral is a *signed* area.

Exercise 3.12: Is $\int_{0}^{3} (x^4 - 5x^3 + 6x^2)\, dx$ positive, negative, or zero?

Make a conjecture about the value of the integral, then check it with technology.

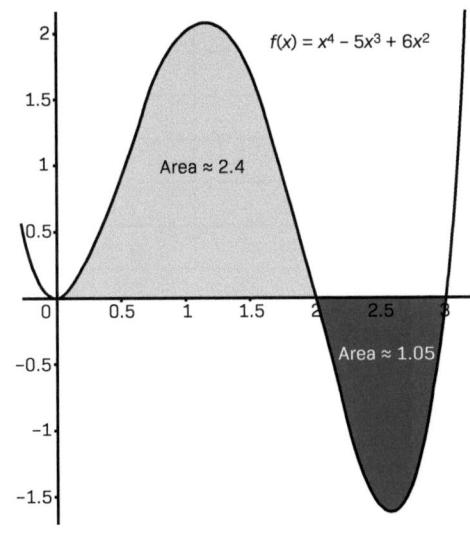

Exercise 3.13: Is $\int_0^{2\pi} \sin(x)\,dx$ positive, negative, or zero?
Make a conjecture about the value of the integral, then check it with technology.

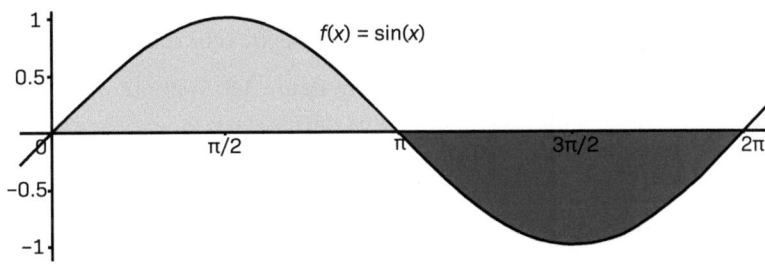

Problem Set 3.2

1. Write out the terms that would be used for a left-hand sum with $n = 5$ that could be used to approximate the definite integral $\int_2^6 \frac{10}{x^2}\,dx$. Please do not evaluate the terms or the sum.

2. A table of values for $g(t)$ is given. Use it to estimate $\int_0^{75} g(t)\,dt$.

t	0	15	30	45	60	75
g(t)	5	7	10	15	37	38

3. Use a calculator or computer to make a table of values of left-hand sums and right-hand sums with 2, 10, 50, 100, and 500 subdivisions. Use these tables to estimate the value of the definite integral. The example of $\int_1^3 (x+2)\,dx$ has been worked for you.

Number of subdivisions	Left-hand sum	Right-hand sum
2	7	9
10	7.8	8.2
50	7.96	8.04
100	7.98	8.02
500	7.996	8.004

Based on the table, I would estimate that $\int_1^3 (x+2)dx = 8$.

a. $\int_2^5 (18-3t)dt$

b. $\int_1^3 \ln(x^2+1)dx$

c. $\int_0^1 e^{x^2} dx$

4. We can write a definite integral to represent the area above the x-axis and under $y = 9 - x^2$.

   a. What are the limits of integration?

   b. What is the value of the definite integral?

5. Consider the integral $\int_0^6 (6-x)dx$.

   a. Interpret the integral as the area of a common shape and find its exact value.

   b. Estimate the integral using technology. Compare your answers to the exact value.

6. Examine the graph of $f(x) = 4 - 2x$, shown in figure 3.8. Without using technology, determine the values of the following definite integrals:

   a. $\int_0^2 f(x)dx$

   b. $\int_{-1}^2 f(x)dx$

   c. $\int_0^4 f(x)dx$

   d. $\int_1^4 f(x)dx$

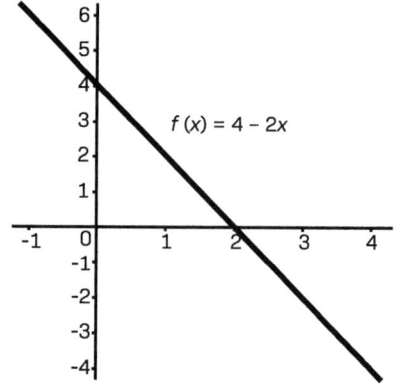

FIGURE 3.8. Graph for problem 6.

7. Let $g(t) = t^2 - 4t + 3$, shown in figure 3.9.
   Without using technology, order the following definite integrals from least to greatest.

$$\int_0^1 g(t)\,dt \quad \int_1^3 g(t)\,dt \quad \int_2^4 g(t)\,dt$$

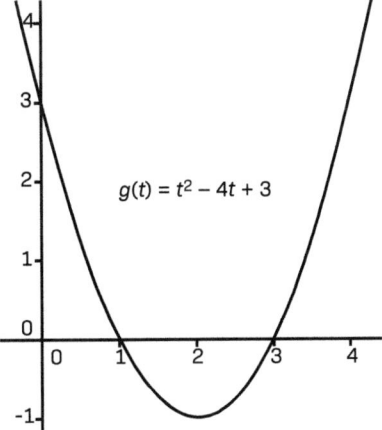

FIGURE 3.9. Graph for problem 7.

8. Suppose that $\int_1^2 h(x)\,dx = -4$ and $\int_1^4 h(x)\,dx = 0$ for the function $h(x)$, shown in figure 3.10. Determine the values of these definite integrals:

$$\int_2^4 h(x)\,dx \quad \int_2^3 h(x)\,dx \quad \int_1^5 h(x)\,dx$$

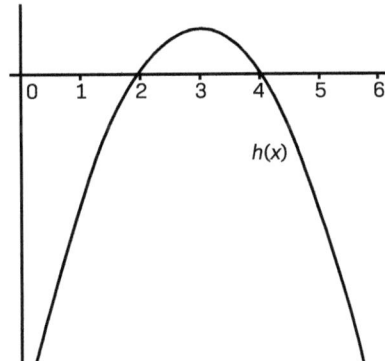

FIGURE 3.10. Graph for problem 8.

3.3 Area Between Curves

OBJECTIVES FOR SECTION 3.3: Upon completing this section, you will be able to do the following:

- Compose and decompose figures to describe the area of a region bounded by two or more functions
- Write expressions with definite integrals to describe the area of a region bounded by two or more functions
- Determine the area of a region bounded by two or more functions

In the last section, we saw that when $f(x)$ is positive on the interval $a \leq x \leq b$ and $a < b$, $\int_a^b f(x)dx$ describes the area that is between the vertical lines $x = a$ and $x = b$ and between the x-axis and the graph of $y = f(x)$.

We can use these ideas to find the area of other regions, such as those that are

- bounded above and below by two different functions, but not the x-axis;
- bounded above by different functions, depending on the values of the domain; and
- bounded by functions that are not always positive.

For example, the following five regions are bounded above by a parabola and below by a straight line.

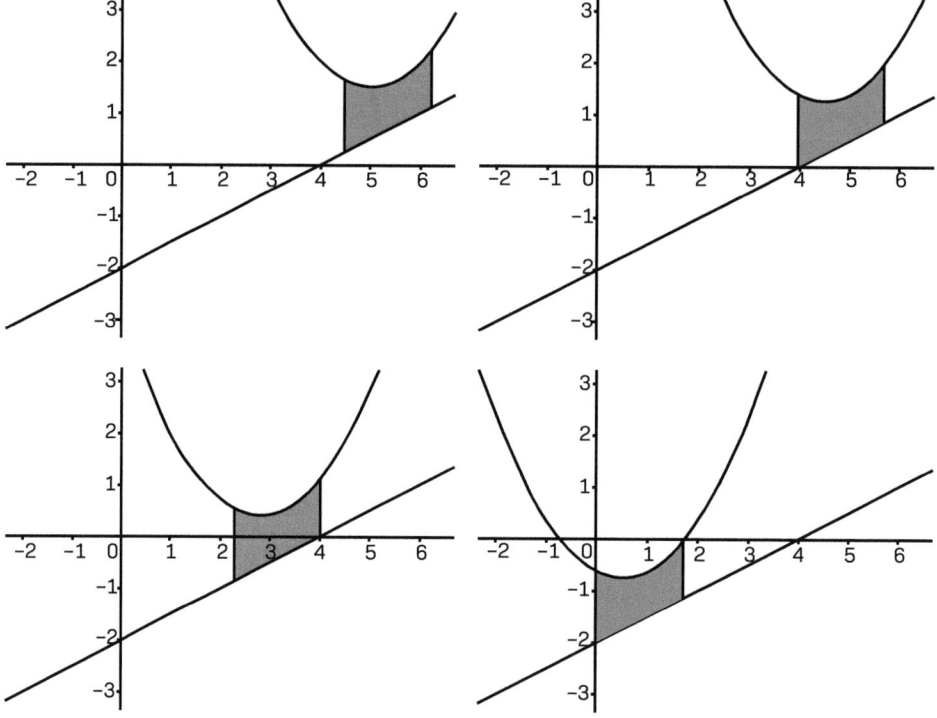

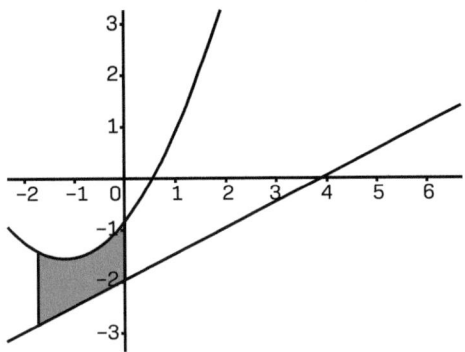

Exercise 3.14: How do the areas of these five regions compare? Are some of them positive, while others are negative? Use the GeoGebra applet "3.3 Area Between Curves" at https://ggbm.at/BN8D2duN to investigate.

Exercise 3.15: Consider Region A, which is bounded by $f(x) = 2x^2$ and $g(x) = 8$ from $x = -2$ to $x = 2$.

a. Sketch the region bounded by the graphs of the functions $y = f(x)$ and $y = g(x)$ on the indicated interval.
b. Write an expression using definite integrals to represent the area of the region.
c. Use technology to estimate the area of the region.

We can use Desmos to graph Region A, as shown in figure 3.11:

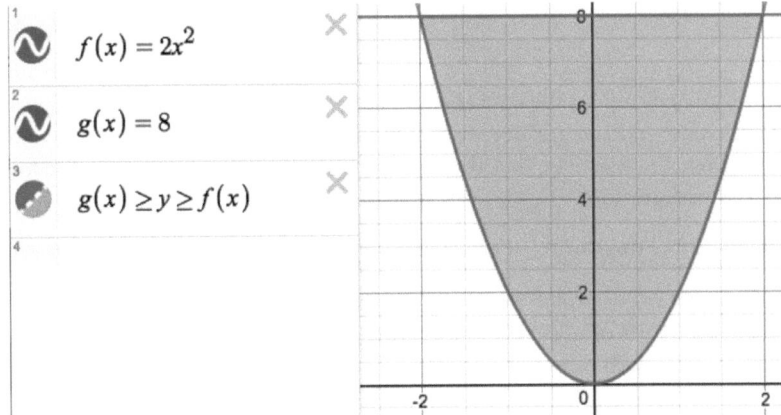

FIGURE 3.11. Graph of Region A for exercise 3.15.

We can imagine this region as being formed by subtracting the area under the parabola from the area under the line, starting at $x = -2$ and ending at $x = 2$. Pictorially, we have what is displayed in figure 3.12:

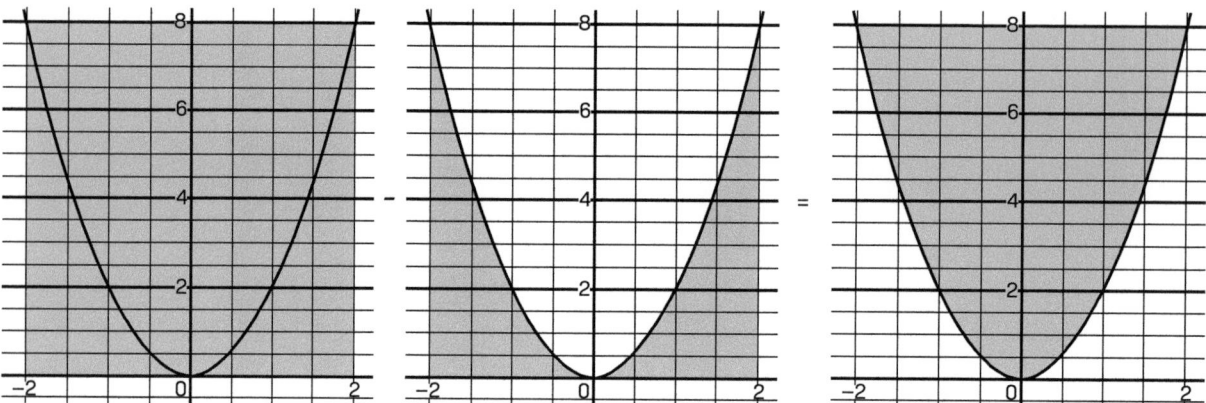

FIGURE 3.12. Creating Region A by decomposing a rectangle.

Box 3.5: Possible Pitfalls

Use pictures to help you see how to find the area of a region through decomposing or composing "familiar" shapes.

Using definite integrals, we can write the area of the region as $\int_{-2}^{2} g(x)dx - \int_{-2}^{2} f(x)dx$

The region under the line is a rectangle with a base of 4 units and a height of 8 units; therefore, the area is 32 square units. According to GeoGebra, the area of the region under the parabola is about 10.67 units. Therefore, the area of Region A is 21.33 square units, as 32 − 10.67 = 21.33.

Exercise 3.16: Repeat this process for Region B, bounded by $f(x) = 13 - 3x^2$ and $g(x) = 1$ from $x = -2$ to $x = 2$.

When we are looking for the area of a region between two functions, GeoGebra allows us to use the command IntegralBetween. It has four inputs: the function that defines the top of the region, the function that defines the bottom of the region, and the lower and upper limits of integration:

IntegralBetween[<Function> , <Function> ,<Start x-Value> , <End x-Value>]

If we apply this command to Region A, we first define the functions **f(x) = 2x²** and **g(x) = 8**. The results of using the IntegralBetween command are shown in figure 3.13. Notice that it agrees with our earlier result.

72 | Fundamentals of Calculus for Teachers

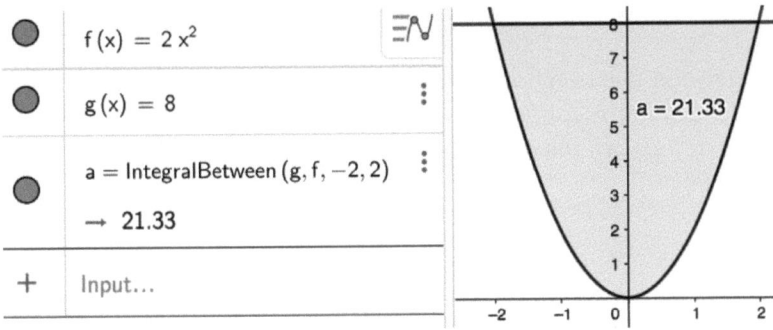

FIGURE 3.13. Using the IntegralBetween command in GeoGebra for Region A.

Exercise 3.17: What would happen if you switched the order of the functions and placed f first in the IntegralBetween command? Why do you suppose that happens?

For Regions A and B, you were given the limits of integration. Sometimes, you will need to determine these on your own. Do so in completing the following exercises with Regions C and D.

Exercise 3.18: Region C is bounded by $f(x) = x^2 + x$ and $g(x) = 3 - x$.

a. Sketch the region bounded by the graphs of the functions $y = f(x)$ and $y = g(x)$ on the indicated interval.
b. Write an expression using definite integrals to represent the area of the region.
c. Use technology to estimate the area of the region.

Box 3.6: Possible Pitfalls

The area of a region should be positive. If you obtain a negative answer, check and see if the definite integrals accurately describe the region.

Exercise 3.19: Region D is bounded by $f(x) = 3x - x^2$ and $g(x) = 4 - 2x$.

a. Sketch the region bounded by the graphs of the functions $y = f(x)$ and $y = g(x)$ on the indicated interval.
b. Write an expression using definite integrals to represent the area of the region.
c. Use technology to estimate the area of the region.

> **Box 3.7: Teaching Tips**
>
> Students (and the teacher!) benefit from discussing different solution methods. Compare your method with a classmate and discuss any differences. If they are the same, see if together you can find a new method of solution.

Exercise 3.20: Write an expression using definite integrals to represent the area of each shaded region in the following graphs.

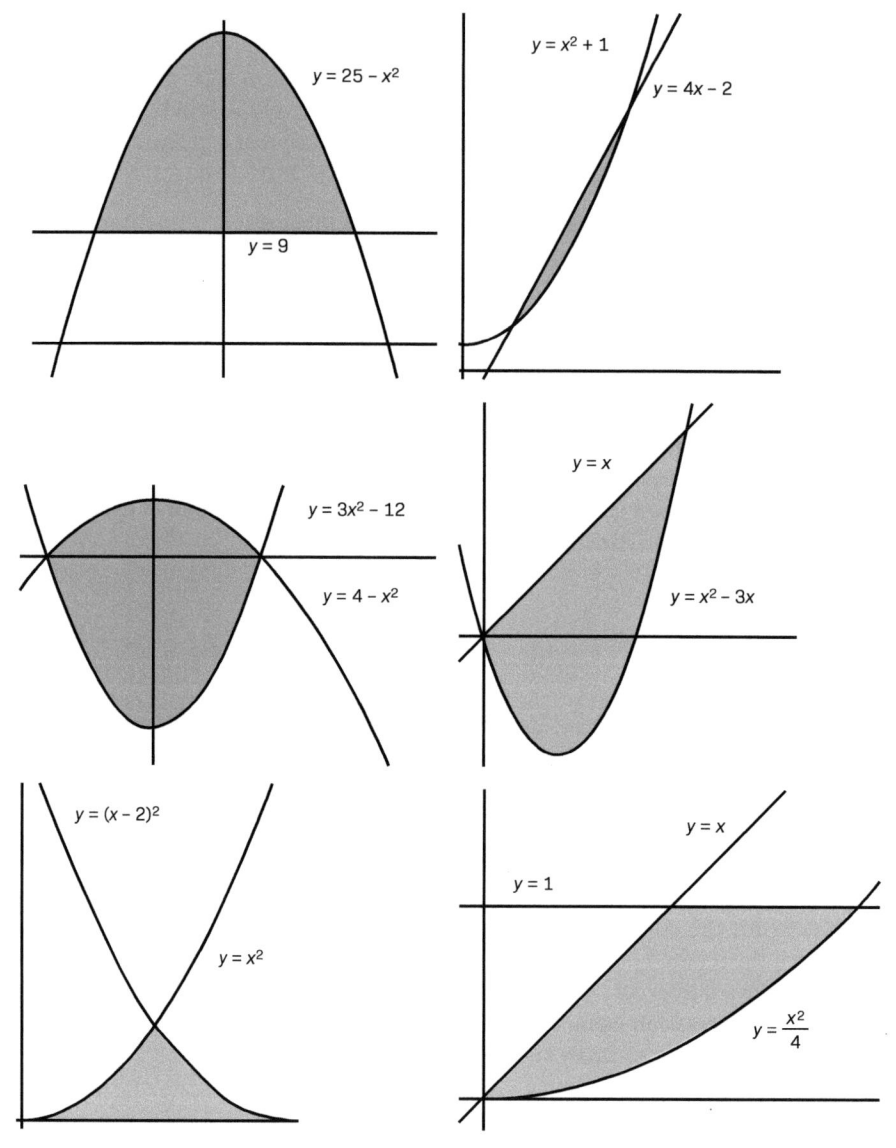

Problem Set 3.3

1. The graph of a function $f(t)$ is shown in figure 3.14. Which of the following numbers could be an estimate of $\int_0^1 f(t)\,dt$ accurate to two decimal places? Explain how you arrived at your answer.

   − 97.35   71.86   96.38   100.12

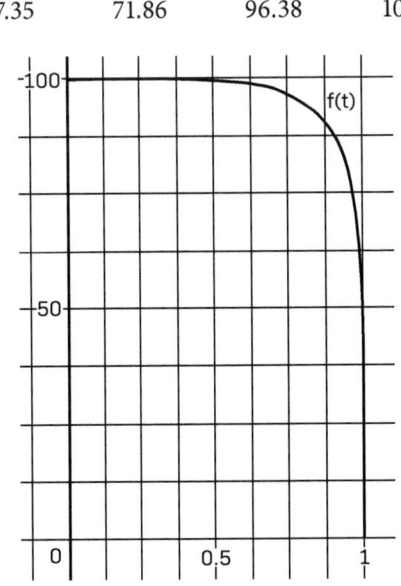

FIGURE 3.14. Graph for problem 1.

2. Examine the graph of the function $j(x)$, shown in figure 3.15:
   a. Find $\int_{-2}^{0} j(x)\,dx$
   b. If the area of the shaded region is represented by the number K, determine the value of $\int_{-2}^{4} j(x)\,dx$.

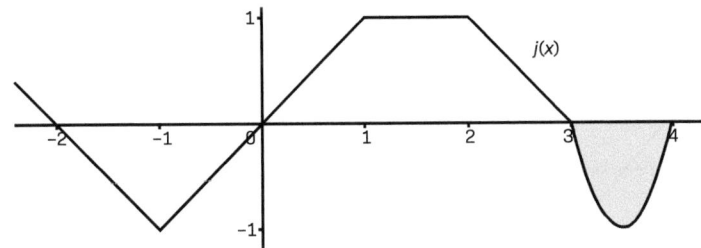

FIGURE 3.15. Graph for problem 2.

For problems 3–7, complete the following three steps:
   a. Sketch the region bounded by the graphs of the functions $y = f(x)$ and $y = g(x)$ on the indicated interval.

b. Write an expression using definite integrals to represent the area of the region.

c. Use technology to determine the area of the region.

3. $f(x) = x(2 - x)$ and $g(x) = 4$ from $x = 0$ to $x = 2$
4. $f(x) = 0.3x^2$ and $g(x) = 0.3x^2 + 1$ from $x = -2$ to $x = 2$
5. $f(x) = \dfrac{1}{x^2}$ and $g(x) = -x^2$ from $x = 1$ to $x = 2$
6. $f(x) = \sqrt{x}$ and $g(x) = 3 - x$ from $x = 0$ to $x = 1$
7. $f(x) = \sqrt{x}$ and $g(x) = 3 - x$ from $x = 2$ to $x = 3$

For problems 8–10, write an expression with definite integrals to represent the area of the region.

8.

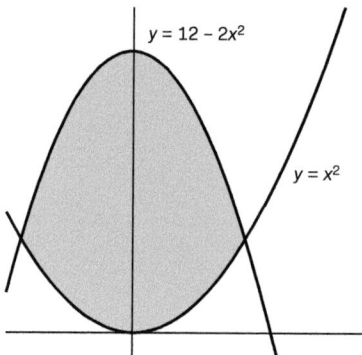

9.

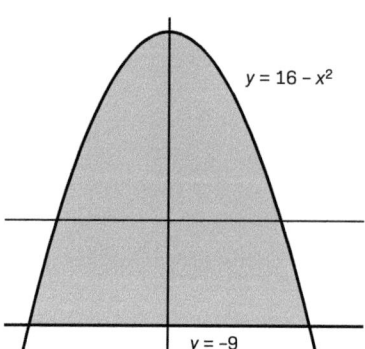

10.

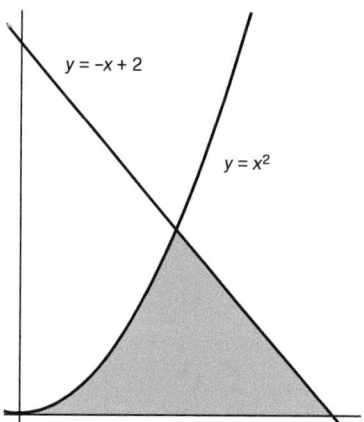

For problems 11–14, complete the following three steps:
a. Sketch the region bounded by the graphs of the functions $y = f(x)$ and $y = g(x)$.
b. Write an expression using definite integrals to represent the area of the region.
c. Use technology to determine the area of the region.

11. $f(x) = x^2$ and $g(x) = 18 - x^2$
12. $f(x) = 2^x$ and $g(x) = \frac{3}{2}x + 1$
13. $f(x) = x^2$ and $g(x) = x^3$
14. $f(x) = x^{1/2}$ and $g(x) = x^{1/3}$

3.4 Interpretations of the Definite Integral

OBJECTIVES FOR SECTION 3.4: Upon completing this section, you will be able to do the following:

- Determine the meaning of the definite integral of a function in a given context
- Determine the units of the definite integral of a function
- Calculate the average value of a function on a given interval
- Write and interpret statements and formulas about the definite integral of a function

In section 2.4, we discussed interpretations of the derivative and the units of the derivative. For example, consider the function $v(t)$ that gives the velocity of an object at time t. Suppose $v(t)$ measured in meters per second, where t is measured in seconds. The derivative $v'(t)$ is the instantaneous rate of change of the velocity, also known as the acceleration. The units of $v'(t)$ are $\frac{\text{meters per second}}{\text{second}}$, or "meters per second squared," a common unit of acceleration.

In this section, we will consider the definite integral. That is, we will address questions such as, "What are the units of $\int_a^b v(t)\,dt$?" and "What is the meaning of $\int_0^5 v(t)\,dt = 4$?"

Recall that we first estimated the value of a definite integral with the sums of areas of rectangles. The heights of those rectangles were obtained from the values of the function, and therefore the units of the heights would be the same as the units of the function—in this case, meters per second. The widths of those rectangles were measured along the horizontal axis and therefore have the same units as the input of the function—in this case, seconds. The area of a rectangle is found by multiplying the height and the width, and therefore the units for the area should be the product of the units for the height and width. In this case, we have the units for the area of each rectangle equal to $\frac{\text{meters}}{\text{second}} \cdot \text{second} = \text{meters}$. The sum of the areas will also be measured in meters.

In general, the units for a definite integral of a function will be the product of the units for the output of the function and the units of the input of the function. The notation of the definite integral can help to remind us that we are finding a sum of products.

$$\int_a^b \underbrace{f(x)}_{\substack{\text{units} \\ \text{of the} \\ \text{output}}} \underbrace{dx}_{\substack{\text{units} \\ \text{of the} \\ \text{input}}} \text{ is a sum with units equal to } \left(\begin{array}{c} \text{units} \\ \text{of the} \\ \text{output} \end{array}\right) \cdot \left(\begin{array}{c} \text{units} \\ \text{of the} \\ \text{input} \end{array}\right)$$

$$\left(\begin{array}{c} \text{units} \\ \text{of the} \\ \text{output} \end{array}\right)\left(\begin{array}{c} \text{units} \\ \text{of the} \\ \text{input} \end{array}\right)$$

Box 3.8: Teaching Tips

When students understand the units of an expression, they are better prepared to interpret the mathematical symbols and numbers. Show your students how to reason about the units within contextual problems.

How, then, should we interpret $\int_0^5 v(t)dt$? The units of the definite integral are meters, and the 0 and 5 both refer to the input, which was measured in seconds. Therefore, $\int_0^5 v(t)dt$ represents the distance traveled by an object between $t = 0$ seconds and $t = 5$ seconds.

Here's another example. Suppose $f(t)$ is the velocity of a runner measured in miles per hour and t is measured in hours since noon. What are the units of $\int_a^b f(t)dt$? How should we interpret $\int_1^3 f(t)dt = 5$?

The units of the definite integral are miles, because that is the product of miles per hour and hours. $\int_1^3 f(t)dt = 5$ means that between 1 p.m. and 3 p.m., the runner is now 5 miles further from where she was at 1 p.m. It could mean that she went 5 miles total, but other possibilities exist, such as she went 7 miles away from home and 2 miles back.

Exercise 3.21: A bathtub is being filled at a rate of $f(t)$ gallons per minute, where t is measured in minutes after 8 p.m.

a. What are the units of $\int_a^b f(t)dt$?

b. What is the meaning of the following definite integrals?

$$\int_0^5 f(t)dt = 100 \qquad \int_{15}^{20} f(t)dt = 0 \qquad \int_{20}^{30} f(t)dt = -80$$

Exercise 3.22: Let $g(x)$ be the cost of making x feet of rope, where $g(x)$ is measured in dollars. What are the units of $\int_a^b g(x)dx$?

Exercise 3.23: Let $p(t)$ be the number of people in a library t hours after midnight.

a. What are the units of $\int_a^b p(t)dt$? What are the units of $\frac{1}{b-a}\int_a^b p(t)dt$?

b. What is the meaning of $\frac{1}{12-8}\int_8^{12} p(t)dt = 100$?

As you can see, we can sometimes get unconventional units for the definite integral. What exactly is meant by the product of feet and dollars? Or the product of people and hours[2]?

The statement $\frac{1}{12-8}\int_8^{12} p(t)dt = 100$ uses a definite integral in a slightly different way. The integral $\int_8^{12} p(t)dt$ has units people hours. Let's interpret the 12 and 8 as hours, so that $\frac{1}{12-8}$ is interpreted as a quantity with units $\frac{1}{\text{hours}}$. Therefore, $\frac{1}{12-8}\int_8^{12} p(t)dt$ has units $\frac{1}{\text{hours}}$ people·hours = people. This means that the number 100 in the equation refers to 100 people. Therefore, we can interpret this equation as "between 8:00 and 12:00, there were an average of 100 people in the library."

This leads to the use of the definite integral in finding the average value of a function. Consider the function $h(t)$, which gives the temperature at the SHSU football stadium. Suppose $h(t)$ is measured in degrees Fahrenheit, and t is measured in hours after midnight on January 1, 2019. If we wanted to find the average temperature at this location on January 1, we could measure the temperature at several times during the day and then calculate the mean by adding the temperatures and dividing by the number of temperatures.

If we measured the temperature every six hours beginning at midnight, our (left-hand sum) estimate for the average temperature would be $\frac{h(0)+h(6)+h(12)+h(18)}{4}$. We could improve this estimate by measuring the temperature every hour, which would result in $\frac{h(0)+h(1)+h(2)+\cdots+h(23)}{24}$. Measuring the temperature every half hour would give this improved estimate: $\frac{h(0)+h(0.5)+h(1)+\cdots+h(23.5)}{48} = \frac{1}{48}(h(0)+h(0.5)+h(1)+\cdots+h(23.5))$

As we improve our estimate, the number of terms in the sum of the numerator increases. We used the variable n to represent this quantity. Additionally, the change in time, Δt, is decreasing. On the interval $a \leq t \leq b$, these quantities are related: $\frac{b-a}{n} = \Delta t$. We can use algebra and divide by $(b-a)$ on both sides to obtain $\frac{1}{n} = \frac{\Delta t}{b-a}$. In the case where temperatures are measured every half-hour, we have $\frac{1}{48} = \frac{0.5}{24-0}$. Therefore, we could manipulate this estimate to appear as

---

[2] Actually, the product people · hours, often written as person-hours, has been used to describe productivity in a workplace. If a task can be completed in 12 person-hours, it could theoretically be completed by one person in twelve hours, two people in six hours, or 24 people in half an hour.

$$\frac{h(0)+h(0.5)+h(1)+\cdots+h(23.5)}{48} = \frac{1}{48}(h(0)+h(0.5)+h(1)+\cdots+h(23.5))$$

$$= \frac{0.5}{24-0}(h(0)+h(0.5)+h(1)+\cdots+h(23.5))$$

$$= \frac{1}{24-0}(h(0)\cdot 0.5+h(0.5)\cdot 0.5+h(1)\cdot 0.5+\cdots+h(23.5)\cdot 0.5)$$

Hopefully, part of this looks familiar. Do you see the Riemann sum on the right-hand side? Inside the parentheses, we have something that looks like $h(t_0) \cdot \Delta t + h(t_1) \cdot \Delta t + \cdots + h(t_{n-1}) \cdot \Delta t$, and the limit of this sum as n approaches infinity is $\int_0^{24} h(t)dt$. Therefore, if we were able to consider an infinite number of temperatures during the day, the average temperature would be $\frac{1}{24-0}\int_0^{24} h(t)dt$.

Exercise 3.24: Consider the function $h(t)$, as described in the previous paragraphs.

a. What are the units of $\frac{1}{24-0}\int_0^{24} h(t)dt$?

b. Write an expression using a definite integral to represent the average temperature at this location between 8 a.m. and 5 p.m.

c. What is the meaning of $\frac{1}{17-9}\int_9^{17} h(t)dt = 40$?

All this leads to a definition of the average value of a function.

Definition 3.4.1: Average Value of a Function

The average value of the continuous function $f(x)$ on the interval $a \leq x \leq b$ is denoted as $\bar{f}$, where $\bar{f} = \frac{1}{b-a}\int_a^b f(x)dx$.

Box 3.9: Teaching Tips

Physical materials, like towers of blocks, can help students understand complex concepts, like the average value of a function.

We can use a graph of the function and make sense of this formula by interpreting the definite integral as an area. One interpretation of the mean is that it "makes even." Consider a set of three towers made from blocks, as in figure 3.16, where the towers have heights of 7 blocks, 6 blocks, and 2 blocks, respectively. The mean, or average, number of blocks used in a tower is $\frac{7+6+2}{3} = 5$. This meaning of "average" implies that if 15 blocks $(7 + 6 + 2 = 15)$ were used to make three towers, they would all be the same height if each tower was 5 blocks tall.

FIGURE 3.16. Finding the average height of towers by making them even.

Use the GeoGebra applet "3.4 Average Value" at https://ggbm.at/BN8D2duN to explore the average value of the function $f(x) = x^2$ on the interval $1 \leq x \leq 4$. Move the slider to find a rectangle that has a base along that interval and an area equal to the area under the function. The height of this rectangle is the average value of the function on that interval.

We can see that the area of the region shown is 21, and therefore $\int_1^4 x^2\,dx = 21$. We are looking to "make the region even" while keeping the area the same. That is, we would like to find a rectangle that has the same area as the region across the same interval.

The interval has a width of 3 units (4 − 1 = 3). Therefore, if the height of this desired rectangle is a number that, when multiplied by 3, gives an area of 21, that means the height (i.e., the average value $\overline{f}$) is 7.

More generally, if the height of the rectangle is $\overline{f}$ and the width is $(b - a)$, then we have
$$\overline{f} \cdot (b-a) = \int_a^b f(x)\,dx, \text{ and therefore } \overline{f} = \frac{1}{b-a}\int_a^b f(x)\,dx.$$

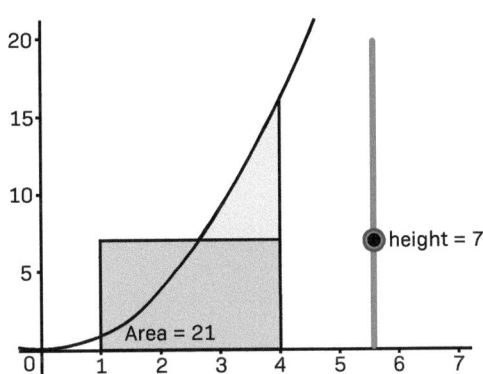

FIGURE 3.17. Using GeoGebra to interpret the average value of a function.

Problem Set 3.4

1. The circumference of a circle centered at (0,0) is $c(r)$, where the radius is r. Both $c(r)$ and r are measured in centimeters.
    a. What are the units of $\int_a^b c(r)\,dr$?
    b. Write sentences to explain the meanings of the following equations.
    $$\int_0^4 c(r)\,dr = 16\pi \qquad \int_2^4 c(r)\,dr = 12\pi$$
    c. Write a definite integral using the function $c(r)$ to represent the shaded area in figure 3.18.

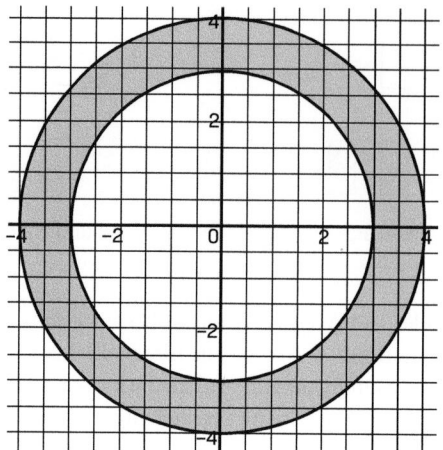

FIGURE 3.18. Shaded area for problem 1.

2. Two cars start at rest, side by side, at a traffic light, then accelerate for several minutes. The graph in figure 3.19 shows their velocities as a function of time. Let $b(t)$ be the velocity of the blue car and $r(t)$ be the velocity of the red car, both measured in feet per minute; t is measured in minutes after the light turns green.

   a. What are the units of $\int_0^2 b(t)\,dt$?
   b. Which car is ahead after one minute? Which car is ahead after two minutes?
   c. Explain how you would estimate the time (after $t = 0$) that the two cars were side by side.

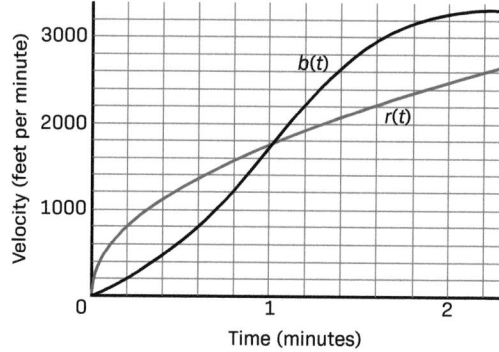

FIGURE 3.19. Velocity versus time graphs for two cars in problem 2.

3. Suppose that $C(t)$ represents the daily cost of heating your house, measured in dollars per day, where t is measured in days and $t = 0$ corresponds to January 1, 2019.
   What is the meaning of the following expressions?

$$\int_0^{90} C(t)\,dt \qquad \int_{31}^{59} C(t)\,dt$$

$$\frac{1}{90-0}\int_0^{90} C(t)\,dt \qquad \frac{1}{59-31}\int_{31}^{59} C(t)\,dt$$

4. Suppose that the population of Angola could be modeled by the function

$$P = f(t) = 10.34(1.032)^t$$

where P is in millions of people and t is in years since 1990. Use this function to predict the average population of Angola between the years 2020 and 2040.

For items 5 and 6, write a definite integral to represent the average value of the function on the given interval. Also, find the average value.

5. Consider the function $p(x) = 8 - 2x$.
    a. Write a definite integral to represent the average value $\bar{p}$ on $0 \leq x \leq 3$.
    b. Determine the average value $\bar{p}$ on $0 \leq x \leq 3$.

6. Consider the function $g(x) = x^2 - 2^x$.
    a. Write a definite integral to represent the average value $\bar{g}$ on $2 \leq x \leq 4$.
    b. Determine the average value $\bar{g}$ on $2 \leq x \leq 4$.
    c. Write a definite integral to represent the average value $\bar{g}$ on $0 \leq x \leq 4$.
    d. Determine the average value $\bar{g}$ on $0 \leq x \leq 4$.

3.5 Theorems About Definite Integrals

OBJECTIVES FOR SECTION 3.5: Upon completing this section, you will be able to do the following:

- Identify relationships between definite integrals with similar integrands or limits of integration
- Explain theorems related to definite integrals
- Use these theorems to combine or simplify definite integrals and determine their values.

3.5.1 Investigations and Explorations

Work with a partner to investigate these sets of definite integrals. Use technology to determine the value of each definite integral and look for patterns.

Box 3.10: Teaching Tips

When students work cooperatively, they gain insight from their peers, and they strengthen their understanding by explaining their own reasoning. Watch how this unfolds as you work with a classmate in this section.

	Partner 1	Partner 2
Set 1	$\int_3^3 x^2 \, dx$	$\int_{-2}^{-2} e^x \, dx$

Exercise 3.25: What do you notice about the values of the definite integrals in set 1? Why does this occur?

	Partner 1	Partner 2
Set 2	$\int_1^3 x^2\, dx$	$\int_3^1 x^2\, dx$
Set 3	$\int_0^{-2} e^x\, dx$	$\int_{-2}^{0} e^x\, dx$
Set 4	$\int_1^0 (-3-x^2)\, dx$	$\int_0^1 (-3-x^2)\, dx$

Exercise 3.26: What do you notice about the values of the definite integrals in sets 2–4? Why does this occur?

	Partner 1	Partner 2
Set 5	$\int_1^3 x^2\, dx,\ \int_3^4 x^2\, dx$	$\int_1^4 x^2\, dx$
Set 6	$\int_{-2}^{5} e^x\, dx$	$\int_{-2}^{0} e^x\, dx,\ \int_0^5 e^x\, dx$

Exercise 3.27: What do you notice about the values of the definite integrals in sets 5 and 6? Why does this occur?

	Partner 1	Partner 2
Set 7	$\int_3^5 x\, dx$	$\int_3^5 10x\, dx$
Set 8	$\int_0^3 7x^2\, dx$	$\int_0^3 x^2\, dx$
Set 9	$\int_{-2}^{5} e^x\, dx$	$\int_{-2}^{5} -3e^x\, dx$

Exercise 3.28: What do you notice about the values of the definite integrals in Sets 7–9? Why does this occur?

	Partner 1	Partner 2
Set 10	$\int_0^3 (x^2+2x)\, dx$	$\int_0^3 x^2\, dx,\ \int_0^3 2x\, dx$
Set 11	$\int_3^5 x\, dx,\ \int_3^5 \ln(x)\, dx$	$\int_3^5 (x-\ln(x))\, dx$

Exercise 3.29: What do you notice about the values of the definite integrals in sets 10 and 11? Why does this occur?

	Partner 1	Partner 2
Set 12	$\int_{4}^{7}(x-4)^2\,dx$	$\int_{0}^{3}x^2\,dx$
Set 13	$\int_{3}^{5}\ln(x)\,dx$	$\int_{13}^{15}\ln(x-10)\,dx$
Set 14	$\int_{-3}^{-1}e^{x+1}\,dx$	$\int_{-2}^{0}e^{x}\,dx$

Exercise 3.30: What do you notice about the values of the definite integrals in Sets 12–14? Why does this occur?

3.5.2 Illustrations and Explanations

In the previous exercises, you identified evidence to support the following six theorems. Following this list, we will discuss why these theorems are true.

> **Theorems about definite integrals**
>
> Suppose $\int_{a}^{b}f(x)\,dx$ and $\int_{a}^{b}g(x)\,dx$ exist.
>
> 3.1 $\int_{a}^{a}f(x)\,dx = 0$ for any real number a
>
> 3.2 $\int_{b}^{a}f(x)\,dx = -\int_{a}^{b}f(x)\,dx$
>
> 3.3 $\int_{a}^{c}f(x)\,dx + \int_{c}^{b}f(x)\,dx = \int_{a}^{b}f(x)\,dx$ for any real number c in the interval from a to b
>
> 3.4 $\int_{a}^{b}[k \cdot f(x)]\,dx = k \cdot \int_{a}^{b}f(x)\,dx$ for any real number k
>
> 3.5 $\int_{a}^{b}[f(x)+g(x)]\,dx = \int_{a}^{b}f(x)\,dx + \int_{a}^{b}g(x)\,dx$
>
> 3.6 $\int_{a+h}^{b+h}f(x-h)\,dx = \int_{a}^{b}f(x)\,dx$ for any real number h.

Theorem 3.1: Going Nowhere

Here's an explanation for the first theorem, that $\int_{b}^{a}f(x)\,dx = 0$ for any real number a. You should have found evidence for this theorem in set 1. In this definite integral, the lower and upper limits of integration are the same number. Therefore, the "region" has a width of zero and has an area equal to zero. Figure 3.20 illustrates this theorem.

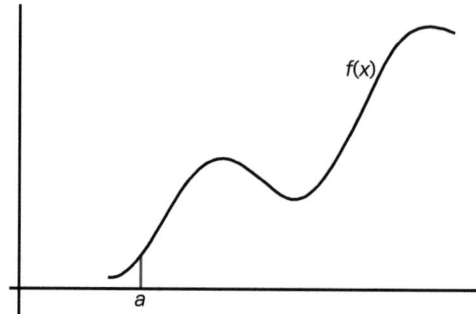

FIGURE 3.20. Illustrating theorem 3.1.

Theorem 3.2: Reversing Direction

Next, let's consider what happens when the lower and upper limits of integration are reversed. According to the second theorem, $\int_b^a f(x)dx = -\int_a^b f(x)dx$. This means that if we reverse the limits of integration, we change the sign of the definite integral. This happens because we are moving in opposite directions when we consider the width of the rectangles used to make approximations. If b is greater than a, then the value $\Delta x = \frac{b-a}{n}$ is positive, as shown in figure 3.21 on the left. If we are examining the area in the opposite direction (as in figure 3.21 on the right), $\Delta x = \frac{b-a}{n}$ will be negative because b is greater than a.

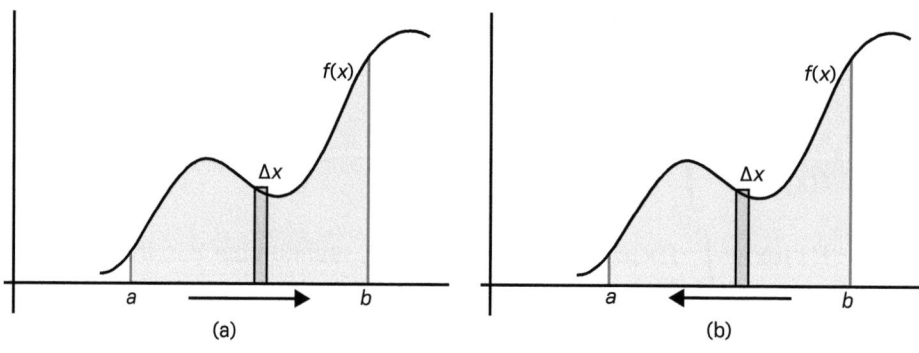

FIGURE 3.21. Illustrating theorem 3.2.

If we compute Riemann sums for both cases—one moving from a to b and the other from b to a—the only difference will be the sign of Δx. Because Δx is a factor of each term in the sum, the two definite integrals will be opposites of each other. If one is positive, then the other will be negative.

Theorem 3.3: Part-Part-Whole

Next, consider the third theorem, where $\int_a^c f(x)dx + \int_c^b f(x)dx = \int_a^b f(x)dx$ for any real number c in the interval from a to b. This theorem is saying that if we split the interval from

a to b into two sub-intervals, the sum of the definite integrals on those two sub-intervals is the definite integral on the entire interval. Figure 3.22 illustrates this idea.

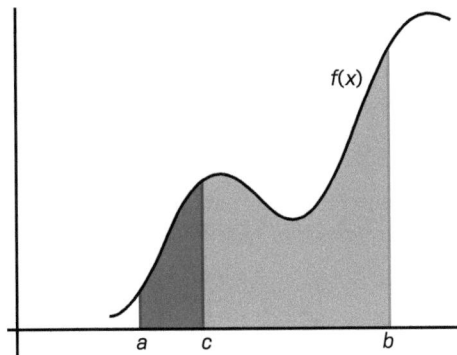

FIGURE 3.22. Illustrating theorem 3.3.

For functions that are continuous over all real numbers, an even stronger statement can be made: $\int_a^c f(x)dx + \int_c^b f(x)dx = \int_a^b f(x)dx$ for *any* real number c, including values of c outside the interval from a to b.

Theorem 3.4: Constant Multiple

The fourth theorem is sometimes referred to as the **constant multiple theorem**. It states that $\int_a^b [k \cdot f(x)]dx = k \cdot \int_a^b f(x)dx$ for any real number k. In terms of transformations of the graph of a function, multiplying the function by a constant k stretches (or compresses) a function vertically by a factor of k. This has the effect of multiplying the heights of the rectangles used in the Riemann sum by that same factor k.

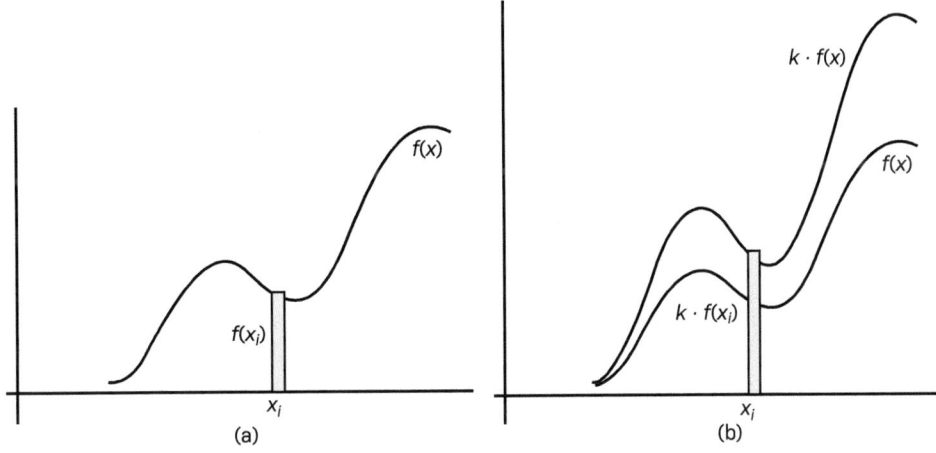

FIGURE 3.23. Part of the Riemann sum of $f(x)$ and $k \cdot f(x)$.

The left-hand Riemann sum for $k \cdot f(x)$ is $k \cdot f(x_0)\Delta x + k \cdot f(x_1)\Delta x + \cdots + k \cdot f(x_{n-1})\Delta x$. We can factor out k to obtain $k(f(x_0)\Delta x + f(x_1)\Delta x + \cdots + f(x_{n-1})\Delta x)$, which is k times the left-hand Riemann sum for $f(x)$. Poetically, we may say that the definite integral of the product of a constant and a function is the product of that constant and the definite integral of the function.

Theorem 3.5: Adding Functions

Even more poetically, the fifth theorem states that definite integral of the sum of two functions is the sum of the definite integrals of those functions: $\int_a^b [f(x)+g(x)]dx = \int_a^b f(x)dx + \int_a^b g(x)dx$. In figure 3.24 on the left, we consider a rectangle from the Riemann sum for $g(x)$ as having a height of $g(x_i)$. The Riemann sum for $f(x) + g(x)$ would consider rectangles that have heights of $f(x_i) + g(x_i)$.

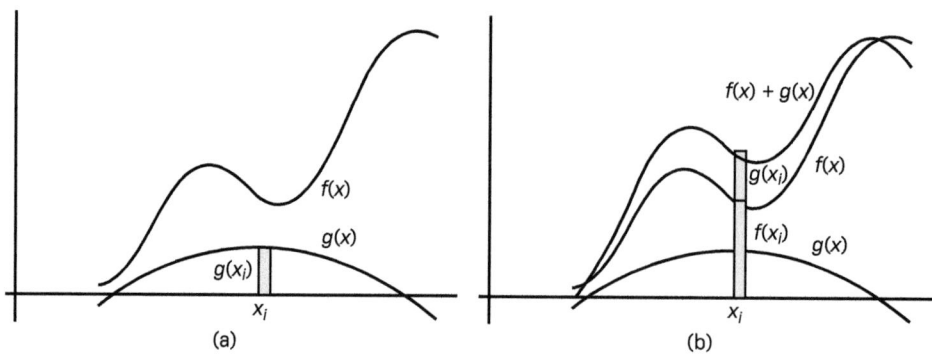

FIGURE 3.24. Part of the Riemann sum of g(x) and f(x) + g(x).

In fact, the region between $g(x)$ and the x-axis on the interval from a to b has the same area as the region between $f(x) + g(x)$ and $f(x)$ on that same interval. Adding g to f essentially "boosts" f up and increases the area accordingly.

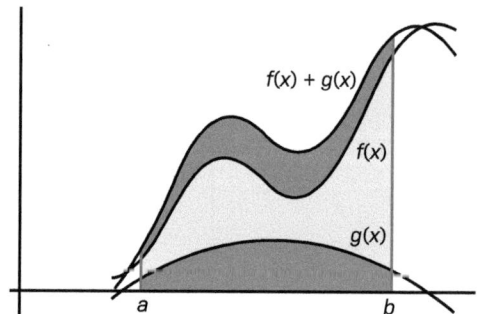

FIGURE 3.25. The function f(x) is boosted up by g(x) to form f(x) + g(x).

A corollary that follows from this theorem relates to the difference of functions: If $\int_a^b f(x)dx$ and $\int_a^b g(x)dx$ exist, then $\int_a^b [f(x)-g(x)]dx = \int_a^b f(x)dx - \int_a^b g(x)dx$.

We may prove this by beginning with $\int_a^b [f(x)-g(x)]dx$, which equals $\int_a^b [f(x)+(-g(x))]dx$. By theorem 3.5, this is equal to $\int_a^b [f(x)dx] + \int_a^b -g(x)dx$. Rewrite this as $\int_a^b f(x)dx + \int_a^b (-1) \cdot g(x)dx$.

This equals $\int_a^b f(x)dx + (-1) \cdot \int_a^b g(x)dx$ by the constant multiple theorem.

Finally, this becomes $\int_a^b f(x)dx - \int_a^b g(x)dx$.

We used the final statement, $\int_a^b f(x)dx - \int_a^b g(x)dx$ in section 3.3 when we examined the area between curves. In fact, this corollary gives us two ways to find the area of a region between two functions: both $\int_a^b f(x)dx - \int_a^b g(x)dx$ and $\int_a^b [f(x) - g(x)]dx$ give the same result. This is particularly useful when we know formulas for the functions and subtracting simplifies things.

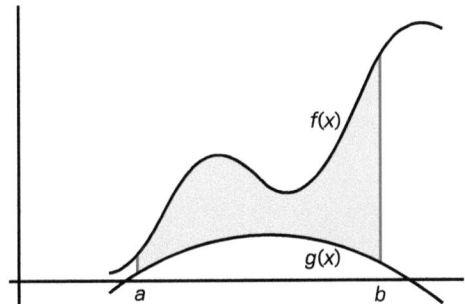

FIGURE 3.26. Applying theorem 3.5 to find the area between curves.

Theorem 3.6: Horizontal Translation

The sixth theorem is sometimes referred to as the **horizontal translation theorem**: If $\int_a^b f(x)dx$ exists, then $\int_{a+h}^{b+h} f(x-h)dx = \int_a^b f(x)dx$ for any real number h. From the perspective of transformations of graphs of functions, the graph of $y = f(x - h)$ can be thought of as a horizontal translation of the graph of $y = f(x)$, h units to the right. Therefore, the limits of integration a and b would also be moved right h units. Translating both the function and the interval of interest horizontally will also translate the regions described in the definite integral. Figure 3.27 shows that these regions have the same area.

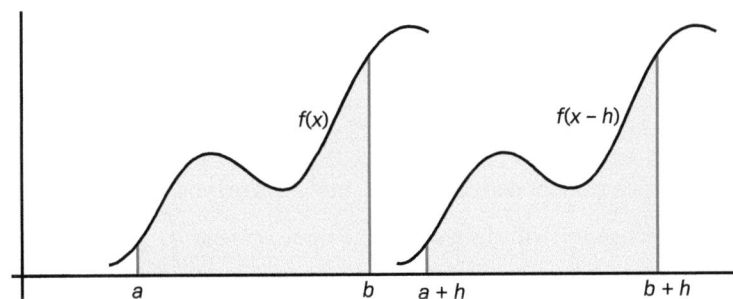

FIGURE 3.27. Illustrating theorem 3.6.

Problem Set 3.5

Suppose that f and g are continuous functions over all real numbers and that

$$\int_1^2 f(x)dx = -4, \int_1^5 f(x)dx = 6, \text{ and } \int_1^5 g(x)dx = 8.$$

Use the theorems from this section to find the value of each integral in problems 1–7.

1. $\int_2^2 g(x)dx$

2. $\int_5^1 g(x)dx$

3. $\int_1^5 3f(x)dx$

4. $\int_4^5 f(x-3)dx$

5. $\int_2^5 f(x)dx$

6. $\int_1^5 [g(x) - f(x)]dx$

7. $\int_1^5 [4f(x) + g(x)]dx$

8. Suppose $\int_2^6 h(x)dx = 3$ and $\int_4^6 h(x)dx = 10$.

   a. Evaluate the following integrals:

   $$\int_2^4 h(x)dx \qquad \int_5^7 h(x-1)dx \qquad \int_0^4 h(x+2)dx$$

   b. Find values for a and k that make the following statement true.

   $$\int_a^{10} k \cdot h(x-4)dx = 25$$

9. Suppose $\int_0^9 q(x)dx = 5$.

   a. Rewrite the expression $\int_0^9 [2 - q(x)]dx + 2\int_9^0 q(x)dx$ as a single definite integral.

   b. Some students are discussing the value of $\int_0^9 [2 - q(x)]dx + 2\int_9^0 q(x)dx$.
   Jailen says the value must be negative.
   Jaidah says the value is 3.
   Ximena says the value is 7.
   Marcelo says the value cannot be determined, even with technology.
   With which student do you agree, if any? Explain your reasoning.

10. Which of the following will always have the same value as $\int_a^b j(x)dx$?

    $0.25\int_a^b 4j(x)dx \qquad \int_{a+4}^{b+4} j(x+4)dx \qquad -\int_b^a j(x)dx \qquad \int_{a+4}^{b+4} [j(x) - 4]dx$

References

Common Core State Standards Initiative. (n.d.). Retrieved from http://www.corestandards.org

Texas Education Agency. (n.d.). Texas education knowledge and skills. Retrieved from https://tea.texas.gov/curriculum/teks/

Credits

Fig. 3.11: Copyright © by Desmos, Inc.
Fig. 3.13: Copyright © by GeoGebra.

Derivatives and Integrals

Applications with Formulas

Think

In the past two chapters, we found instantaneous velocity by approximating the slope of a curve describing distance as a function of time, and we found net distance traveled by approximating the area under a curve describing the velocity as a function of time. Both of these situations involve the same quantities . . . so how are they related? Before we address that question, we'll uncover some patterns that will allow us to find formulas for the derivatives of many types of functions. Knowing these patterns will help us reveal the relationship between derivatives and integrals.

Remember

To prepare for this chapter, you may want to think about the following topics. They serve as prerequisites and primers to the mathematical content in the chapter. Some of these ideas come from chapters 2 and 3.

- Use the rules of exponents to write expressions in different ways, such as $\sqrt{x} = x^{1/2}$, $\frac{1}{x^2} = x^{-2}$, and $\sqrt[4]{x^5} = x^{1.25}$.
- Know the basic shape and properties of an exponential function, such as $j(x) = 3 \cdot 2^x$.
- Think of examples of functions that have inverses, such as $f(x) = \ln(x)$ and $f^{-1}(x) = e^x$, or $g(x) = x^3$ and $g^{-1}(x) = \sqrt[3]{x}$.
- Write a formula for the composition of two functions, such $f(f^{-1}(x))$, $f(g(x))$, and $g(f(x))$ using functions such as $f(x) = \ln(x)$ and $g(x) = x^3$.

Connect

Across all grade levels, students work to develop proficiency in mathematical practices, and eight mathematical practices are described in the Common Core State Standards for Mathematics (CCSS-M) (n.d.). In this chapter, we will "look for and make use of structure" (CCSS-M Practice MP7) and "look for and express regularity in repeated reasoning" (CCSS-M Practice MP8) as we develop formulas for the derivatives of functions and as we uncover a relationship between derivatives and integrals.

4.1 Derivatives of Powers and Polynomials

OBJECTIVES FOR SECTION 4.1: Upon completing this section, you will be able to do the following:

- Understand theorems related to the derivative of a constant multiple or sums of functions
- Find derivatives of power functions and polynomial functions
- Apply your knowledge to find the second and third derivative (and beyond) of a function

In chapter 2, we investigated the concept of the derivative of a function. In this chapter, we will further explore the derivatives of particular types of functions. This section is devoted to power functions and polynomial functions. In the next two sections, we will learn about derivatives of exponential and logarithmic functions. Overall, we will look for patterns and learn how the formula for a function can give us information about its derivative.

We will begin by recalling the definition of the derivative function. After that, we will state and prove a few theorems about derivatives that will be helpful in our investigation.

> **Definition 4.1.1: The Derivative Function**
>
> The derivative of a function $f(x)$ is defined as $f'(x) = \lim\limits_{h \to 0} \left(\dfrac{f(x+h) - f(x)}{h} \right)$, provided that this limit exists. (This is also Definition 2.3.1.)

If a function g is a constant multiple of a function f, how are their derivatives g' and f' related? Let's investigate this question using Desmos.

Exercise 4.1: Begin with the function $f(x) = x(x - 2)^2 - 1$ in Desmos.

Next, type $g(x) = c \cdot f(x)$ and add a slider for c. By moving the slider for c, you can transform the graph of f with a vertical stretch or compression by a factor of c.

Compare the derivatives of these two functions by typing $f'(x)$ and $g'(x)$ on separate lines. How do the derivatives change when c changes?

Note that moving the slider also changes the graph of the derivative of g. In fact, it appears that one derivative graph is the vertical stretch or compression of the first, using the same scale factor. That is, it appears that $g'(x) = c \cdot f'(x)$. (You can get even more evidence for this conjecture by graphing $c \cdot f'(x)$ with Desmos.)

While Desmos provides evidence that this claim, "If $g(x) = c \cdot f(x)$, then $g'(x) = c \cdot f'(x)$" is true, it does not actually *prove* it true for all cases. To prove it, we need to use the definition of the derivative. What follows is a proof that the claim is indeed true. In fact, I will call it a theorem.

Box 4.1: Possible Pitfalls

Resist the temptation to simply memorize the theorem and skip over the proof. Seek to understand *why* the theorem is true.

Theorem 4.1

If $g(x) = c \cdot f(x)$, then $g'(x) = c \cdot f'(x)$.

Proof: Suppose $f(x)$ is a function and that its derivative $f'(x)$ exists. If $g(x) = c \cdot f(x)$ for some real number c, by the definition of the derivative, $g'(x) = \lim_{h \to 0} \left(\frac{g(x+h) - g(x)}{h} \right)$. Because $g(x) = c \cdot f(x)$, we have $g'(x) = \lim_{h \to 0} \left(\frac{c \cdot f(x+h) - c \cdot f(x)}{h} \right)$. Because c is a constant, we may factor it out of the numerator, and even out of the limit. Thus, $g'(x) = c \cdot \lim_{h \to 0} \left(\frac{f(x+h) - f(x)}{h} \right)$. Finally, the limit statement that remains is the definition of the derivative of f, so we have $g'(x) = c \cdot f'(x)$.

Another helpful theorem deals with the derivative of the sum of two functions.

Theorem 4.2

If $j(x) = f(x) + g(x)$, then $j'(x) = f'(x) + g'(x)$.

Proof: Suppose $f(x)$ and $g(x)$ are functions and that their derivatives $f'(x)$ and $g'(x)$ exist. If $j(x) = f(x) + g(x)$, by the definition of the derivative, $j'(x) = \lim_{h \to 0} \left(\frac{j(x+h) - j(x)}{h} \right)$. Because $j(x) = f(x) + g(x)$, we have

$$j'(x) = \lim_{h \to 0} \left(\frac{[f(x+h) + g(x+h)] - [f(x) + g(x)]}{h} \right)$$

$$= \lim_{h \to 0} \left(\frac{f(x+h) + g(x+h) - f(x) - g(x)}{h} \right)$$

$$= \lim_{h \to 0} \left(\frac{f(x+h) - f(x) + g(x+h) - g(x)}{h} \right)$$

$$= \lim_{h \to 0} \left(\frac{f(x+h) - f(x)}{h} + \frac{g(x+h) - g(x)}{h} \right)$$

$$= \lim_{h \to 0} \left(\frac{f(x+h) - f(x)}{h} \right) + \lim_{h \to 0} \left(\frac{g(x+h) - g(x)}{h} \right)$$

$$= f'(x) + g'(x)$$

A simple corollary to this theorem describes the derivative of the difference of two functions.

Corollary 4.3

If $j(x) = f(x) - g(x)$, then $j'(x) = f'(x) - g'(x)$.

Exercise 4.2: In chapter 2, we understood that the derivative could be interpreted as the slope of a function. Use this idea to determine the derivatives of the following functions:

$$f(x) = 3 \qquad g(x) = -9 \qquad k(x) = 2x + 4 \qquad p(x) = -x$$

In this section, we will investigate power functions and polynomial functions. Let's start with the definition of a power function.

Definition 4.1.2: Power Function

A power function can be written in the form $f(x) = k \cdot x^n$, where k and n are real numbers.

Exercise 4.3: Here are some examples of power functions. For each example, state the values of k and n.

$$g(x) = 5x^2 \qquad h(x) = \sqrt{3} \cdot x^7 \qquad j(x) = -9x^{3.4} \qquad l(x) = 144$$

$$p(x) = 0.2\sqrt{x} \qquad q(x) = \frac{1}{x^2} \qquad r(x) = \frac{7}{\sqrt[5]{x^4}} \qquad t(x) = 2\pi x$$

Theorem 4.1 says that if we know the derivative of a function $f(x) = x^n$, then the derivative of $g(x) = k \cdot x^n$ will be k times the derivative of $f(x)$. Using our notation, we may succinctly state $g'(x) = k \cdot f'(x)$. Therefore, it would be nice if we had a way to know the derivatives of functions of the form $f(x) = x^n$.

Let's begin making a list of functions and their derivatives. We will continue to use, and add to, this list throughout the chapter.

Function $f(x) =$	Derivative $f'(x) =$
1	
c (c is a constant)	
x	
mx (m is a constant)	
mx + b (m and b are constants)	
x^2	
x^3	
x^4	
x^5	
⋮	
x^n	

The first five functions on this list are linear, so it should be fairly straightforward to find their derivatives. The quadratic function $f(x) = x^2$ provides us with a bit of a challenge. In section 2.3, we used the definition of the derivative to show that if $f(x) = x^2$, then $f'(x) = 2x$. That work is briefly reproduced as follows:

$$f'(x) = \lim_{h \to 0} \left(\frac{f(x+h) - f(x)}{h} \right) = \lim_{h \to 0} \left(\frac{(x+h)^2 - x^2}{h} \right) = \lim_{h \to 0} \left(\frac{x^2 + 2xh + h^2 - x^2}{h} \right)$$

$$= \lim_{h \to 0} \left(\frac{2xh + h^2}{h} \right) = \lim_{h \to 0} \left(\frac{h(2x+h)}{h} \right) = \lim_{h \to 0} (2x + h) = 2x + 0 = 2x$$

What about $f(x) = x^3$? Again, we use the definition of the derivative. It is a bit more work, because we have to expand $(x + h)^3$, which is equal to $x^3 + 3x^2h + 3xh^2 + h^3$.

$$f'(x) = \lim_{h \to 0} \left(\frac{f(x+h) - f(x)}{h} \right) = \lim_{h \to 0} \left(\frac{(x+h)^3 - x^3}{h} \right) = \lim_{h \to 0} \left(\frac{x^3 + 3x^2h + 3xh^2 + h^3 - x^3}{h} \right)$$

$$= \lim_{h \to 0} \left(\frac{3x^2h + 3xh^2 + h^3}{h} \right) = \lim_{h \to 0} \left(\frac{h(3x^2 + 3xh + h^2)}{h} \right) = \lim_{h \to 0} \left(3x^2 + 3xh + h^2 \right)$$

$$= 3x^2 + 3x(0) + 0^2 = 3x^2$$

Therefore, if $f(x) = x^3$, then $f'(x) = 3x^2$. We could repeat this process for higher powers of x, but let's check the list that we have made thus far.

Exercise 4.4: Do you notice any patterns in the table? If so, make a conjecture for the derivatives of the functions further down the table. You may use GeoGebra to check your conjectures using the command **derivative[<Function>]**.

Function $f(x) =$	Derivative $f'(x) =$
1	0
x	1
x^2	$2x$
x^3	$3x^2$
x^4	
x^5	
$\vdots$	
x^n	

Hopefully, you noticed a pattern that leads to the *power rule* for derivatives.

Power rule for derivatives: If $f(x) = x^n$, then $f(x) = nx^{n-1}$.

Box 4.2: Teaching Tips

Don't be angry that you are just now learning this. Be thankful that you understand the *meaning* of the derivative and you now know a simple way to find the formula for the derivative function. When you are a teacher, help your students learn concepts before they practice procedures.

Exercise 4.5: Does the power rule hold for linear functions like $g(x) = x$ and $h(x) = 1$? If so, why? If not, why not?

The power rule is a marvelous shortcut. It even works when the powers are not natural numbers.

Exercise 4.6: Use the power rule and theorem 4.1 to find the derivatives of these functions.

$$f(x) = x^7 \qquad h(x) = \sqrt{3} \cdot x^7 \qquad g(x) = 5x^2 \qquad k(x) = -2x^3$$

$$j(x) = x^{3.4} \qquad l(x) = -9x^{3.4} \qquad p(x) = \sqrt{x} \qquad p(x) = 0.2\sqrt{x}$$

$$u(x) = x^{-1} \qquad u(x) = \pi x^{-1} \qquad q(x) = \frac{1}{x^2} \qquad r(x) = \frac{7}{x^2}$$

Polynomial functions can be written as the sum of power functions, where the powers are all whole numbers. Therefore, $f(x) = x^7 + \sqrt{3} \cdot x^4 - 0.8x - 3$ is a polynomial function. On the other hand, $u(x) = x^{-1}$, $l(x) = -9x^{3.4}$, and $g(x) = 5x^2 - \sqrt[3]{x}$ are not polynomial functions. Armed with the power rule and the theorems from this section, we are able to find the derivatives of polynomial functions and also the sums (and differences) of any power functions—even if they are not polynomials.

Exercise 4.7: Find formulas for the derivatives of these functions.

$$f(x) = x^7 + \sqrt{3} \cdot x^4 - 0.8x - 3$$
$$g(x) = 5x^2 - \sqrt[3]{x}$$
$$h(x) = x^{99} - 0.1x^{-2} + 8 + \sqrt{\pi} \cdot x$$
$$j(x) = 5\sqrt{x} - \frac{10}{x^3} + \frac{1}{2\sqrt{x}}$$

After you complete your work, check your answers with GeoGebra. You may find some cases where the GeoGebra output looks different but is still equivalent to your answer.

With these types of functions, it is not very difficult to find the second derivative. We simply apply the power rule and theorems to the first derivative. Furthermore, we can find the third derivative, fourth derivative, and so on, if we desire.

Exercise 4.8: Find the first, second, third, and fourth derivatives of the following function:

$$f(t) = 3t^4 + 5t^8 - 20t$$
$$f'(t) =$$
$$f''(t) =$$
$$f'''(t) =$$
$$f''''(t) =$$

As we discussed in chapter 2, if the function $f(t)$ refers to the position of an object at time t, then the first derivative $f'(t)$ will give the velocity at time t, and the second derivative $f''(t)$ will give the acceleration at time t. The third derivative would describe the rate of change of acceleration with respect to time. Some physicists have named this quantity the *jerk*. In general, it appears to be difficult to conceptualize meanings for higher derivatives, and there is some debate about their meanings, names,[1] and practical usefulness.

Problem Set 4.1

For problems 1–4, let $f(x) = 2x - 8$ and $g(x) = -3x + 6$.

1. a. Let $k(x) = f(x) + g(x)$. Write a formula for $k(x)$ in terms of x.
   b. Verify that $k'(x) = f'(x) + g'(x)$.

2. a. Let $h(x) = 2f(x) - 3g(x)$. Write a formula for $h(x)$ in terms of x.
   b. Verify that $h'(x) = 2f'(x) - 3g'(x)$.

3. a. Let $j(x) = f(g(x))$. Write a formula for $j(x)$ in terms of x.
   b. Find a formula for the derivative of $j(x)$. Were you expecting this result?

4. a. Let $p(x) = f(x) \cdot g(x)$. Write a formula for $p(x)$ in terms of x.
   b. Is it true that $p'(x) = f'(x) \cdot g'(x)$?

Find the derivatives of the given functions in problems 5–13.

5. $y = x^9$
6. $y = -8x^{-7}$
7. $y = x^{2.3}$
8. $y = x^{7/4}$
9. $y = \dfrac{1}{x}$
10. $y = x^{3/8}$
11. $y = 3t^5 - 4\sqrt{t} + \dfrac{3}{t^2}$
12. $y = 5t^3 + \dfrac{4}{t} + \pi\sqrt[3]{t^2} + 1$
13. $y = \pi x + ex^2 + e$

Use GeoGebra (www.geogebra.org/classic) to find the derivatives of the functions listed in problems 14–16. For each function, state whether the power rule applies.

14. $y = 3^x$
15. $y = x^\pi - x^{-\pi}$
16. $y = \dfrac{1}{x^2 + 1}$

17. Let $f(t) = 2t^3 - 4t^2 + 3t - 1$.
    a. Find formulas for $f'(t)$ and $f''(t)$.
    b. Use your formula for $f''(t)$ to determine when the graph of f is concave up.

---

[1] The fourth derivative of position with respect to time has been called both the *jounce* and *snap*. (http://en.wikipedia.org/wiki/Jounce)

18. Let $g(x) = x^6 + x^5 + x^4 + x^3 + x^2 + x^1 + 1$. Write formulas for the first, second, third, fourth, fifth, sixth, seventh, and eighth derivatives of g.

19. A grapefruit is thrown into the air. The equation $s = 6 + 100t - 16t^2$ describes the height of the grapefruit above the ground (in feet) after t seconds.

    a. Find a formula for the velocity of the grapefruit.

    b. When is the velocity zero?

    c. Find a formula for the acceleration of the grapefruit.

    d. How do you explain the acceleration you found in part c?

20. Prove Corollary 4.3: If $j(x) = f(x) - g(x)$, then $j'(x) = f'(x) - g'(x)$.

4.2 Derivatives of Exponential Functions

OBJECTIVES FOR SECTION 4.2: Upon completing this section, you will be able to do the following:

- Find derivatives of exponential functions
- Define the number e and the number $\ln(b)$ for $b > 0$
- Solve problems involving derivatives of exponential functions

In this section, we will investigate the derivatives of exponential functions. In general, exponential functions have the form $f(x) = a \cdot b^x$, where a is a real number and b is a positive real number. Let's begin by using Desmos to graph examples of a special kind of exponential function where $a = 1$.

Box 4.3: Possible Pitfalls

As you saw in problem 14 of problem set 4.1, if $y = 3^x$, then $\frac{dy}{dx}$ is not $x3^{x-1}$. The power rule doesn't apply to exponential functions.

Exercise 4.9: Begin by typing $f(x) = b^x$ and then add a slider for b. Next, graph the derivative function by typing $f'(x)$. In figure 4.1, I used a dotted line for the original function and a solid line for the derivative.

Move the slider for b to examine several different exponential functions. Notice that the graph does not appear for negative values of b.

What do you notice about the derivative when $0 < b < 1$?

What do you notice about the derivative when $b > 1$?

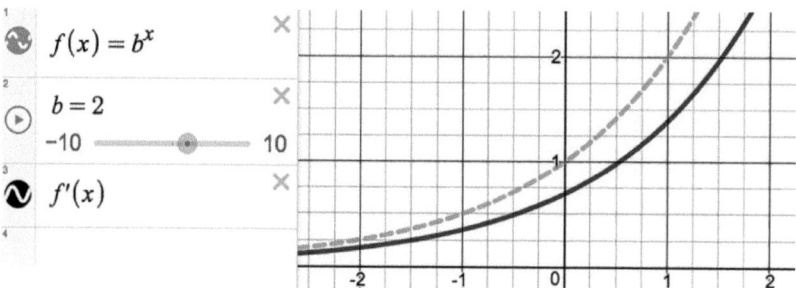

FIGURE 4.1. Using Desmos to graph the derivative of an exponential function.

When b is between 0 and 1, the original function is decreasing, and the derivative is (appropriately) always negative. When b is greater than 1, the original function is increasing, and the derivative is always positive.

When the derivative exists, it looks quite a bit like an exponential function itself. If you try, you may be able to see that the graph of the derivative appears to involve a vertical stretch (or compression) of the original function. The scale factor of the vertical stretch appears to depend on the value of b. In the case shown in figure 4.1, where $b = 2$, the scale factor is less than 1, and the derivative graph is compressed toward the x-axis. In a case such as $b = 6$ in figure 4.2, the scale factor appears to be greater than 1.

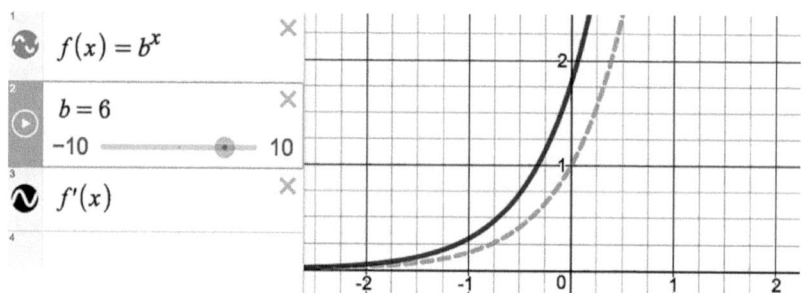

FIGURE 4.2. The derivative of $f(x) = 6^x$ is a vertical stretch of the original function.

Using our knowledge of transformations of functions, it appears that if $f(x) = b^x$, then $f'(x) = k \cdot f(x)$, or $f'(x) = k \cdot b^x$, where k is the scale factor of the vertical stretch. This leads us to two questions that we will investigate:

1. How do we find k if we know the value of b?
2. Is there a value of b where the function and derivative are the same? In other words, is there a special b such that $f'(x) = f(x) = b^x$?

To investigate the first question, let's continue to use Desmos to estimate the scale factors of these vertical stretches and compressions.

We can see that the y-intercept of the original function, $f(x) = b^x$, is always equal to 1. In other words, $f(0) = 1$ for all $b > 0$. Therefore, if $f'(x) = k \cdot f(x)$, then $f'(0) = k \cdot f(0) = k \cdot 1 = k$. This means that the y-intecept of the derivative function is the scale factor k that we seek.

Exercise 4.10: Complete the table by estimating k (the y-intercept of the derivative) for the given values of b in the function $f(x) = b^x$. Note that Desmos will provide the coordinates of the y-intercept, if you click on that location. Write your estimates to at least two decimal places.

b	0.1	0.3	0.5	0.7	0.9	1	2	3	4	5	6
k											

Create a table in Desmos to plot the points (b, k) from the table.
What type of function appears to fit these points? (It's not linear or quadratic.... What other types are there?) Test your conjecture with Desmos.

We haven't quite finished answering the first question, but I can't resist thinking about the second question. To get a function equal to its derivative, we would need $f'(x) = f(x) = b^x$. This would mean that $k = 1$. I notice that none of the examples in the table from exercise 4.10 answer this question. When b is 2, k is less than 1. When b is 3, k is greater than 1. Therefore, I have a guess that this special value of b is somewhere between 2 and 3.

Now, back to the first question. How do we find k if we know b? Let's return to the definition of the derivative, and use the function $f(x) = b^x$.

$$f'(x) = \lim_{h \to 0} \left(\frac{f(x+h) - f(x)}{h} \right) = \lim_{h \to 0} \left(\frac{b^{x+h} - b^x}{h} \right)$$

Notice $b^{x+h} = b^x \cdot b^h$ by the laws of exponents. Therefore, we may write

$$f'(x) = \lim_{h \to 0} \left(\frac{b^x \cdot b^h - b^x}{h} \right).$$

In the numerator, we can factor out b^x. Because neither b nor x depend on h, we can bring them outside the limit. Symbolically, this is written as follows:

$$f'(x) = \lim_{h \to 0} \left(\frac{b^x (b^h - 1)}{h} \right) = b^x \cdot \lim_{h \to 0} \left(\frac{b^h - 1}{h} \right)$$

Notice here that we have written the derivative as the product of the original function b^x and a number (in the form of a limit statement) that depends on b. This limit statement is precisely the value of k for which we have been searching. When you were investigating with Desmos earlier, you should have found that if $f(x) = b^x$, then the scale factor k is $\ln(b)$, and $f'(x) = b^x \cdot \ln(b)$. We can combine this information with the limit previously shown to get a definition of $\ln(b)$.

Definition 4.2.1: The Natural Logarithm of b

$$\ln(b) = \lim_{h \to 0} \left(\frac{b^h - 1}{h} \right) \text{ for } b > 0$$

Therefore, our first question is answered. If $f(x) = b^x$, then $k = \ln(b)$ and $f'(x) = b^x \cdot \ln(b)$. It is probably a good idea to add this function and its derivative to the list we began in section 4.1.

We know $\ln(x)$ as the natural logarithm function. This will help us answer the second question: Is there a value of b such that $f'(x) = f(x) = b^x$? For this to happen, we would need k to be equal to 1. Because $k = \ln(b)$, we need a value of b where $\ln(b) = 1$. To get rid of the logarithms, we could rewrite this expression using exponents. The number e is the base of the natural logarithm, which means $b = e^1$, or simply $b = e$. Notice that $e = 2.718281828459...$, which puts it between 2 and 3 (as we suspected earlier).

This is *really fantastic news*. It means that if $f(x) = e^x$, then $f'(x) = e^x$. The derivative of e^x is e^x. We should add this to the list of functions and derivatives. If e has ever caused you trouble in the past, this should make up for it. It is a key component to one of the simplest derivative formulas out there.

We can use the definition of the natural logarithm to get a definition for the number e. We combine the fact that $\ln(e) = 1$ with the definition $\ln(e) = \lim_{h \to 0}\left(\frac{e^h - 1}{h}\right)$ and get $\lim_{h \to 0}\left(\frac{e^h - 1}{h}\right) = 1$.

This means that for values of h that are close to 0, $\frac{e^h - 1}{h}$ is close to 1. I will write this as $\frac{e^h - 1}{h} \approx 1$ when h is close to 0. If we treat this like an equation, we could multiply both sides by h to get $e^h - 1 \approx h$, and therefore $e^h \approx 1 + h$, when h is close to 0. To "solve" for e, we raise both sides to the power of $\frac{1}{h}$ (or take the hth root) to get $e \approx (1+h)^{\frac{1}{h}}$ when h is close to 0. This work provides us with the following definition of e.

Definition 4.2.2: The Number e

$$e = \lim_{h \to 0}\left((1+h)^{\frac{1}{h}}\right)$$

Box 4.4: Possible Pitfalls

Remember that e is a constant, while e^x and x^e are functions of x.

Exercise 4.11: Use your knowledge of the derivatives of exponential functions, along with theorems 4.1 and 4.2, to find formulas for the derivatives of these functions.

$$f(x) = 2 \cdot 3^x + 5e^x \qquad g(x) = e^x + x^e + 2e + e^2$$

$$h(x) = (0.5)^x \qquad j(x) = 1.3^x$$

$$v(t) = \left(\frac{1}{3}\right)^t \qquad w(t) = \frac{1}{\ln(2)} \cdot 8^t$$

Use GeoGebra (I recommend https://www.geogebra.org/classic) to check your answers. Explain any differences that you notice when you compare your answers to the output from GeoGebra.

Problem Set 4.2

Find the derivatives of the given functions in exercises 1–9.
1. $y = 5^x$
2. $y = 4 \cdot 3^x - 6$
3. $y = 15 \cdot 0.65^x$
4. $y = 3e^x + x^3$
5. $f(t) = 8t^2 + 5e^t$
6. $y = 10e^x + 11^x$
7. $g(x) = 4x^5 + 2e^x + \ln(3)$
8. $f(\theta) = \pi^\theta + \theta + e^\theta$
9. $y = x - e + x^e - e^x$

Use GeoGebra (www.geogebra.org/classic) to find the derivatives of the functions listed in exercises 10–15.

For each function, state whether the rules we have studied for finding derivatives apply.

10. $y = \sqrt{x} + \left(\dfrac{1}{2}\right)^x$
11. $y = x^2 \cdot 2^x$
12. $y = \dfrac{e^x}{x}$
13. $y = x \cdot e^x$
14. $y = (3^x - 5)^\pi$
15. $y = e^{\sqrt{5x+3}}$

16. The population of Latvia can be described by the formula $P = 2.66(0.988)^t$, where P is measured in millions of people and t is measured in years since the start of 1990.

    a. At what rate was the population changing on January 1, 2018?

    b. Explain why the units for the answer to this question are "millions of people per year."

17. The video "Gangnam Style" debuted on July 15, 2012 (officialpsy, 2012). By August, the video had "gone viral." As of September 12, 2012, the total number of views grew to 150 million (YouTube Trends, 2012).

    We can use the function $T(d) = 30(1.038)^d$ to model this relationship. The units of T are millions of views and the units of d are days with $d = 1$ corresponding to August 1, 2012.

    a. Find a formula for the rate at which the total number of views is increasing.

    b. At what rate were the total number of views increasing on August 31, 2012? Please include units with your answer.

    c. Do you think this formula describes the total number of views to the present day? (As of July 2018, this video had over 3.1 billion views.)

18. Let $g(x) = e^x + 2^x + x^2$. Write formulas for the first, second, third, and fourth derivatives of g.

19. Use derivatives to explain why the graph of $y = e^x$ is increasing and concave up for all x.

4.3 Derivatives of Composite Functions

OBJECTIVES FOR SECTION 4.3: Upon completing this section, you will be able to do the following:

- Write a complicated function as the composition of two simpler functions
- Find the derivative of a composite function using formulas and tables
- Find the derivative of the natural logarithm function, $\ln(x)$

Now we know how to find the derivatives of power functions, exponential functions, and constant multiples and additive combinations of these functions. On the other hand, we have not yet discussed how to find the derivatives of functions such as $y = (3^x - 5)^\pi$ or $y = e^{\sqrt{5x+3}}$. These are both examples of composite functions, which we will address in this section. Next is a context that uses a composite function.

Exercise 4.12: The length, L, of a column of mercury in a particular thermometer is a function of the temperature, T. That is, $L = f(T)$. In this example, L is measured in millimeters and T is measured in °C.

Also, the temperature is a function of the time of day, x, measured in hours since midnight. We will say $T = g(x)$.

We could also think of the length as a function of time, where $L = h(x)$ and the function h is a composition of f and g. In symbols, $h(x) = f(g(x))$.

Use the information that follows to fill in the missing information in the equations.

Suppose that length of the column of mercury changes at a rate of 2 mm per °C, so $\dfrac{d}{d} =$

At 8 a.m., the temperature was changing at a rate of 1.7 °C per hour, which means $\dfrac{d}{d} =$

Therefore, at 8 a.m., we also know $\dfrac{d}{d} =$

How fast is the length of the column of mercury changing at 8 a.m.?

Here's the same question: What is $\dfrac{dL}{dx}$ at 8 a.m.?

Here's the same question: What is $h'(8)$?

It is sometimes helpful to break a complicated function into some simpler functions. To do so, we may introduce another variable that links the input and the output. Watch the video of the "floating egg" experiment (HooplaKidzLab, 2013) at http://youtube.com/watch?v=zszw6uCiQpc. Here, the depth of the floating egg (y) is a function of the amount of salt in the water (*x*). A linking variable could be the density of the liquid (*z*). We could set up functions to describe the relationships as follows:

$y = h(x)$ The depth of the egg is a function of the amount of salt in the water.

$y = f(z)$ The depth of the egg is a function of the density of the liquid.

$z = g(x)$ The density of the liquid is a function of the amount of salt in the water.

Notice that the output of *g* is the input of *f*. Therefore, we may compose these two functions to say that $y = f(g(x))$. This composition has the same input and output as *h*, which leads us to writing the single function *h* as a composition of *f* with *g*: $h(x) = f(g(x))$.

A small change in the amount of salt should lead to a small change in the density of the liquid, which will in turn lead to a small change in the depth of the egg. These small changes should remind us of the derivative. How is the derivative of *h* related to the derivatives of *f* and *g*?

Using some alternative notation, we have $h'(x) = \frac{dy}{dx}$, $f'(z) = \frac{dy}{dz}$, and $g'(x) = \frac{dz}{dx}$.

The notation suggests that if we multiply $\frac{dy}{dz}$ and $\frac{dz}{dx}$, the *dz* would be eliminated.

Therefore, it appears that $h'(x) = \frac{dy}{dx} = \frac{dy}{dz} \cdot \frac{dz}{dx} = f'(z) \cdot g'(x)$.

This equation is more simply written as $h'(x) = f'(z) \cdot g'(x)$, which involves both the variables *z* and *x*. To eliminate a variable, recall that $z = g(x)$, so $h'(x) = f'(g(x)) \cdot g'(x)$.

Here are two more examples to help you practice examining composite functions in context.

Exercise 4.13: Suppose that in the dairy industry, the cost of producing a gallon of milk (*m*, in dollars per gallon of milk) is a function of the cost of cattle feed (*c*, in dollars per pound).

Write a formula for this function. Call it *f*.

Additionally, the sale price of ice cream (*i*, in dollars per gallon of ice cream) is a function of the cost of producing a gallon of milk.

Write a formula for this function. Call it *g*.

Write a composite function to describe this scenario. Call it *h*.

Write a formula for the derivative of *h* in terms of *f* and *g* and their derivatives.

What are the units of *h*′?

Exercise 4.14: A spherical balloon is being inflated. For a sphere with radius *r*, the surface area is $S = 4\pi r^2$. At a certain point in time, the radius is 15 cm and is increasing at 0.5 cm per min. How fast is the surface area changing at that time? (Remember to give units.)

To find the derivative of a composite function, it is helpful to decompose it into some simpler functions. The process of finding the derivative of a composite function is sometimes called the *chain rule*.

The Chain Rule

If $z = g(x)$ and $h(x) = f(z) = f(g(x))$, then $h'(x) = f'(z) \cdot g'(x)$.

We may put this entirely in terms of x: If $h(x) = f(g(x))$, then $h'(x) = f'(g(x)) \cdot g'(x)$.

In practice, I like to first identify the functions that could be used to make the composite function. Consider the example $y = (3^x - 5)^\pi$. We could decompose this into two functions, which I will refer to as the *inside function* and the *outside function*. If $h(x) = f(g(x))$, I would call g the inside function and f the outside function.

In this particular case, where $y = (3^x - 5)^\pi$, I will say that the formula for the inside function is $z = g(x) = 3^x - 5$. If this is the case, then the formula for the outside function would be $y = f(z) = z^\pi$. This choice was made so that $y = h(x) = f(g(x)) = f(3^x - 5) = (3^x - 5)^\pi$.

For the chain rule, we need the derivatives of f and g: $g'(x) = 3^x \cdot \ln(3)$ and $f'(z) = \pi z^{\pi-1}$. We now apply the chain rule to find the derivative of $y = h(x) = f(g(x))$.

$$y' = h'(x) = f'(z) \cdot g'(x)$$
$$= \pi z^{\pi-1} \cdot 3^x \cdot \ln(3)$$
$$= \pi (3^x - 5)^{\pi-1} \cdot 3^x \cdot \ln(3)$$

I like to use the following format to organize my work in finding the derivative of a composite function. The derivative looks pretty scary, but I plan to leave it that way.

$h(x) = (3^x - 5)^\pi$

	Outside	Inside (z)
Function	z^π	$3^x - 5$
Derivative	$\pi z^{\pi-1}$	$3^x \cdot \ln(3)$

$$h'(x) = \pi z^{\pi-1} \cdot 3^x \cdot \ln(3)$$
$$= \pi (3^x - 5)^{\pi-1} \cdot 3^x \cdot \ln(3)$$

Box 4.5: Teaching Tips

When you are teaching, it is important to organize your work. It helps to explain this organization to your students. Here, I organized my work to help identify the functions and derivatives I use in this process.

Exercise 4.15: Think of each of the following as a composition of two functions. Write the formulas of the functions that were composed to make this function. Then find the derivative of the composite function.

Example: $f(x) = (2x - 1)^3$

	Outside	Inside (z)
Function	z^3	$2x - 1$
Derivative	$3z^2$	2

$$f'(x) = 3z^2 \cdot 2 = 3(2x - 1)^2 \cdot 2 = 6(2x - 1)^2$$

a. $f(x) = (x^2 + 1)^{100}$
b. $g(x) = 4^{\sqrt{x}}$
c. $h(x) = \dfrac{1}{x^2 + x^4}$
d. $j(x) = \sqrt[5]{e^x + x^2 + 1}$
e. $k(x) = e^{x^2}$
f. $m(x) = 2^{\sqrt{e^x - x}}$

In some cases, we can find the derivative of a composite function with tables of values.

x	f(x)
1	1
2	3
3	2

x	f'(x)
1	3
2	0
3	-3

x	g(x)
1	1
2	3
3	5

x	g'(x)
1	1.5
2	2
3	2.5

If $h(x) = f(g(x))$, here's how we can find $h'(2)$ and $h'(3)$.
By the chain rule, $h'(x) = f'(g(x)) \cdot g'(x)$.
Therefore, $h'(2) = f'(g(2)) \cdot g'(2)$. Using the tables, we have $h'(2) = f'(3) \cdot g'(2) = -3 \cdot 2 = -6$.
Let's try the same process to find $h'(3)$. We start with $h'(3) = f'(g(3)) \cdot g'(3) = f'(5) \cdot g'(3)$. Unfortunately, we have a problem here: We don't know the value of $f'(5)$. Therefore, we say $h'(3)$ is undefined.

Exercise 4.16: Use the following tables to find the derivatives, if possible:

x	f(x)
1	1
2	3
3	2

x	f'(x)
1	9
2	-1
3	10

x	g(x)
1	7
2	13
3	11

x	g'(x)
1	0
2	12
3	5

If $h(x) = f(g(x))$, find $h'(2)$.
If $j(x) = g(f(x))$, find $j'(2)$.
If $p(x) = f(f(x))$, find $p'(2)$.

One excellent application of the chain rule is finding the derivative of an inverse function. Recall that if f and g are inverse functions, then we have $f(g(x)) = x$ for all x in the domain of g, and also $g(f(x)) = x$ for all x in the domain of f.

Some examples of pairs of inverse functions are listed. Notice that in each case, $f(g(x)) = x$ and $g(f(x)) = x$.

- $f(x) = x + 5$ and $g(x) = x - 5$
- $f(x) = 7x$ and $g(x) = \dfrac{x}{7}$
- $f(x) = x^3$ and $g(x) = x^{\frac{1}{3}}$
- $f(x) = e^x$ and $g(x) = \ln(x)$

We know how to find formulas for the derivatives of most of the functions listed. Let's use our knowledge of the chain rule to find a derivative for the natural logarithm function.

Let $f(x) = e^x$ and $g(x) = \ln(x)$. We want to find a formula for $g'(x)$.

These functions are inverses of each other, so $f(g(x)) = x$.

Take the derivative of both sides of this equation to obtain $f'(g(x)) \cdot g'(x) = 1$.

Notice that $f'(x) = e^x$, so $f'(g(x)) = e^{g(x)} = e^{\ln(x)} = x$. This gives us $x \cdot g'(x) = 1$.

Solving for $g'(x)$, we have $g'(x) = \dfrac{1}{x}$.

We should add this to our list of derivatives. The derivative of $\ln(x)$ is $\dfrac{1}{x}$. That's a pretty simple-looking derivative for a complicated function.

Problem Set 4.3

Find the derivative of each function in problems 1–14.

1. $f(x) = (3x + 4)^3$
2. $f(x) = (x^2 + 3x + 4)^5$
3. $f(x) = \dfrac{1}{2 + 3x^4}$
4. $f(x) = \dfrac{1}{(1 - x^2)^4}$
5. $f(x) = \sqrt{x^2 + 3}$
6. $f(x) = 5^{x^2 - 2x + 3}$
7. $f(x) = \sqrt{e^x - 1}$
8. $f(x) = e^{-2x}$
9. $f(x) = \ln(x^2 - 5x + 6)$
10. $f(x) = 4^{\ln(x)}$
11. $f(x) = \ln(\ln(x))$
12. $f(x) = e^{e^x + x}$
13. $f(x) = \sqrt{2^{x^5 - 4} + 3}$
14. $f(x) = \ln(e^{3x-1} + 2)$

Use the following tables for problems 15–18.

x	f(x)
1	3
2	3.6
3	1
4	15

x	f'(x)
1	4
2	3
3	2
4	1

x	g(x)
1	1.5
2	3
3	2.5
4	1

x	g'(x)
1	4
2	5
3	9
4	11

If $h(x) = f(g(x))$, find the following:

15. $h'(2)$
16. $h'(4)$

If $j(x) = g(f(x))$, find the following:

17. $j'(1)$
18. $j'(3)$

4.4 Antiderivatives

OBJECTIVES FOR SECTION 4.4: Upon completing this section, you will be able to do the following:

- Graph multiple antiderivatives of a function, given the graph of the function
- Understand how different antiderivatives of a function are related to each other
- Understand the inverse relationship of derivatives and antiderivatives

In section 3.1, we discussed how to measure the distance traveled using rates of change. Here is a scenario along those lines:

> Denise works in a tall building. She steps into the elevator and travels down at 8 meters per second for 5 seconds, pauses for 2 seconds, and then travels up at 6 meters per second for 3 seconds.

Figure 4.3 displays a graph of Denise's vertical velocity with respect to time. In real life, it would probably be unpleasant to ride this elevator, with its abrupt changes in velocity.

Exercise 4.17: There are some questions we *can* answer in this scenario, and there are other questions that we *cannot* answer. Look over the list that follows and classify the questions as to whether we "can" or "cannot" answer them.

a. After 10 seconds, is Denise above or below where she started?
b. At what height did Denise start?

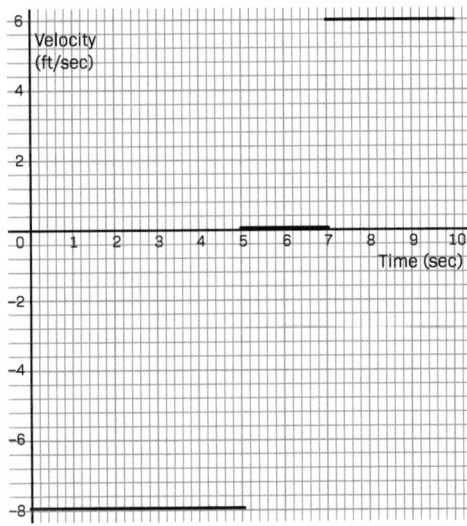

FIGURE 4.3. Denise's velocity in an elevator.

c. At what height does Denise finish, after 10 seconds?
d. What is the net distance that Denise travels in the first second?
e. What is the net distance that Denise travels in the first 5 seconds?
f. What is the net distance that Denise travels in the first 7 seconds?
g. What is the net distance that Denise travels in the first 10 seconds?
h. When does Denise travel below the ground floor?

As you look over these questions, you may realize that some of them deal with the change in Denise's position (a, d, e, f, and g), while others deal with her actual position (b, c, h). Because we don't know anything about where Denise starts or finishes, we cannot answer b, c, or h at this point, but we can answer the items dealing with the change in her position.

For example, Denise travels 8 meters down in the first second, because she was moving at −8 meters per second for one second (d). In the first 5 seconds, her net distance traveled is 40 meters down (e). Because she doesn't move during seconds 5 and 7, her net distance traveled in the first seven seconds is also 40 meters down (f). During the final three seconds, she travels up 18 meters, the product of 6 meters per second and 3 seconds. Therefore, her net distance is 22 meters down from where she started, because −40 + 18 = −22 (g and a).

Exercise 4.18: Suppose we had some additional information about Denise. I'll call it Version I.

Version I. Denise begins 50 meters above the ground.

With this information, we can now answer all the questions. Make a table and draw a graph of Denise's height above the ground during these 10 seconds. A blank graph is provided in Figure 4.4.

TIME (SEC)	0	1	2	3	4	5	6	7	8	9	10
VERSION I HEIGHT (M)											

Exercise 4.19: We could also create tables and graphs for some other versions of this story.

Version II. Denise begins 24 meters above the ground.
Version III. Denise is at the ground level at six seconds.
Version IV. Denise is at the ground level at ten seconds.

TIME (SEC)	0	1	2	3	4	5	6	7	8	9	10
VERSION II HEIGHT (M)											
VERSION III HEIGHT (M)											
VERSION IV HEIGHT (M)											

What do you notice about the graphs Denise's height in these versions? How are they related?

> **Box 4.6: Possible Pitfalls**
>
> Add graphs for Versions II, III, and IV to your graph of Version I to facilitate comparisons.

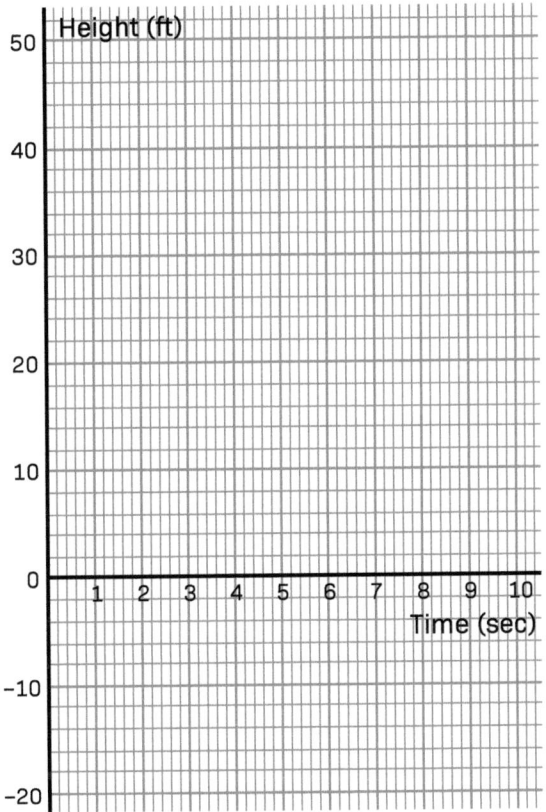

FIGURE 4.4: Graph of height v. time for Versions I–IV.

With a derivative, we begin with a function and determine the rate of change. In this scenario, we are working in the opposite direction. We began with a rate of change and came up with some different versions of the function that had this rate of change. Depending on the conditions in Versions I–IV, we obtained different functions describing her position. Therefore, we say that these functions describing Denise's position are all *antiderivatives* of the function describing her velocity.

All of these antiderivatives are related in the following ways:

1. They all have the same derivative.
2. The graphs of the antiderivatives are vertical shifts of one another.

There is actually no limit to the number of different antiderivates of a function. However, once we identify a specific point on the antiderivative (as in Versions I–IV), we do obtain a single antiderivative.

The process of finding an antiderivative from a graph of $y = f(t)$ involves the definite integral. In the elevator scenario with Denise, let's use $f(t)$ to represent her velocity at time t. To find the net distance that Denise traveled in the first second, we use $\int_0^1 f(t)dt$. Similarly, $\int_0^{10} f(t)dt$ gives us the net distance that she traveled in the first 10 seconds.

We often use a capital letter to represent an antiderivative of a function. For example, an antiderivative of a function f would be F, and G could be used to refer to an antiderivative of g.

Definition 4.4.1: Antiderivative

If F is an antiderivative of f, then $F'(t) = f(t)$.

This means the derivative of an antiderivative of f is f. In this sense, the derivative and the antiderivative "undo" each other.

A function has many antiderivatives, but they are all related. If F_1 and F_2 are both antiderivatives of f, then there is a real number C such that $F_1(t) = F_2(t) + C$.

The definite integral gives us a net change in the function. If we know the value of an antiderivative at a particular input, we can find the value of the antiderivative at other points.

For example, consider Version I of the scenario. Let's use $F(t)$ to represent Denise's position at time t, so that we have $F(0) = 50$. Because $\int_0^5 f(t)dt$ gives us the change in position in the first 5 seconds, the position at $t = 5$ seconds will be $F(5) = F(0) + \int_0^5 f(t)dt = 50 + (-40) = 10$

With Version 2, $F(0) = 24$, and $F(5) = F(0) + \int_0^5 f(t)dt = 24 + (-40) = -16$.

In general, we can say that $F(b) = F(a) + \int_a^b f(t)dt$. The position at time b is equal to the position at time a plus the net change in position from time a to time b.

We can take this equation, $F(b) = F(a) + \int_a^b f(t)dt$, and manipulate it algebraically to obtain $\int_a^b f(t)dt = F(b) - F(a)$. This important result is part of the Fundamental Theorem of Calculus, which we will discuss in the next section.

Box 4.7: Possible Pitfalls

Remember that the area of a triangle is one-half the product of the base and height, and the definite integral gives a signed area.

Exercise 4.20: The two graphs that follow represent functions.

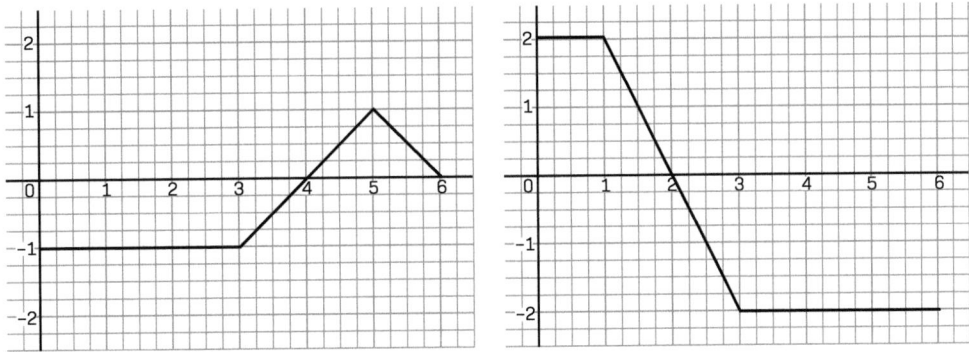

On the grids below these graphs, sketch *three* different antiderivatives for each function.

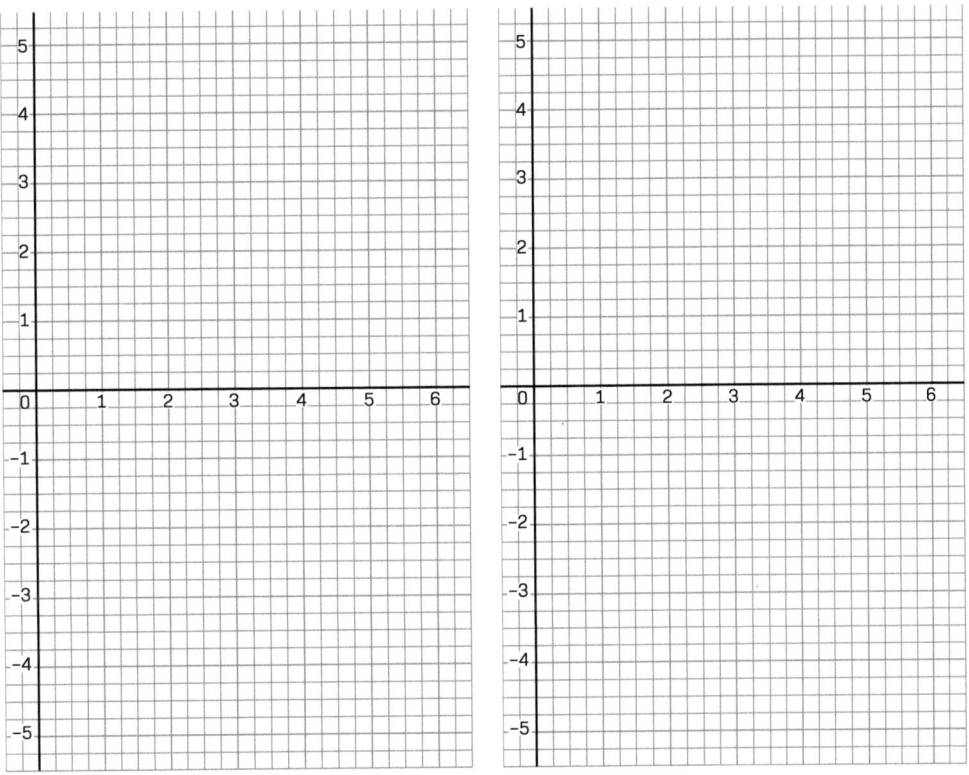

Exercise 4.21: The two graphs that follow represent functions.

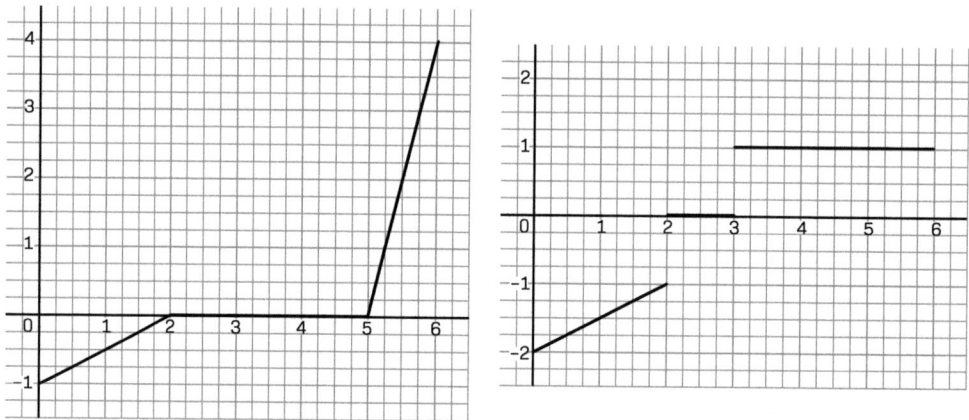

On the grids below these graphs, sketch three different antiderivatives for each function.

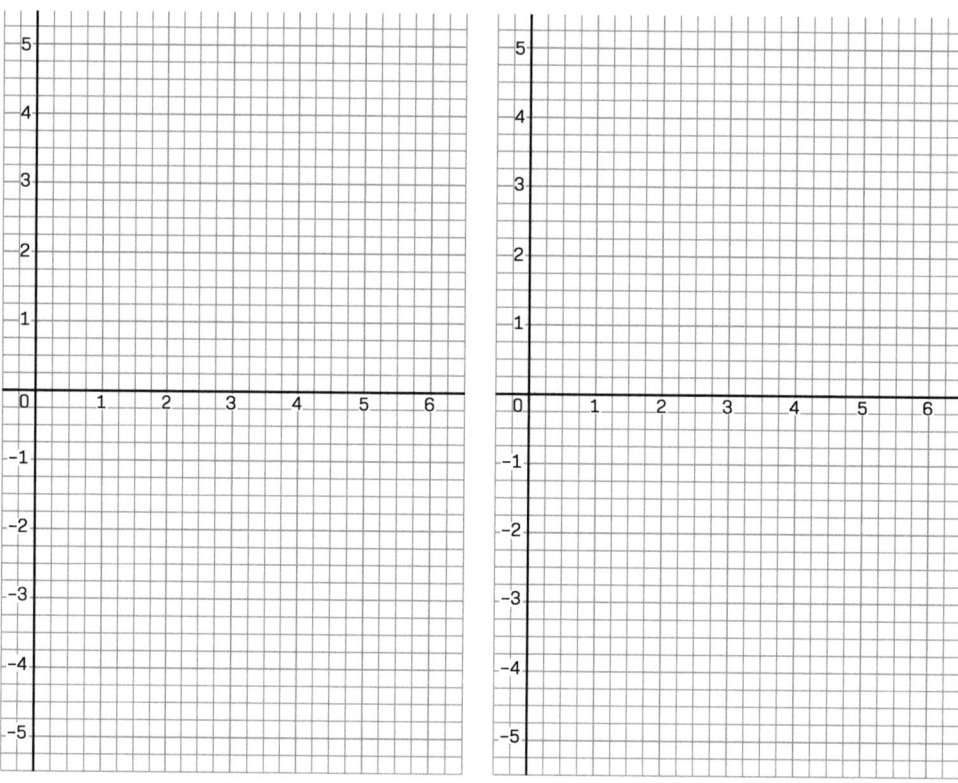

Problem Set 4.4

The graphs in problems 1–6 represent functions. For each function, draw graphs of *two* different antiderivatives. Make one of the antiderivatives pass through the origin, (0,0).

1.

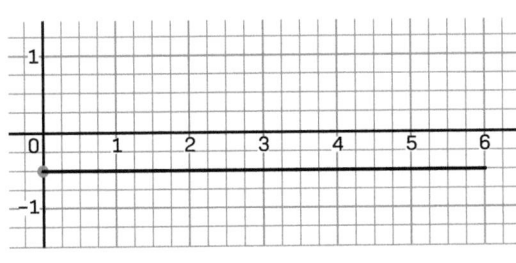

2.

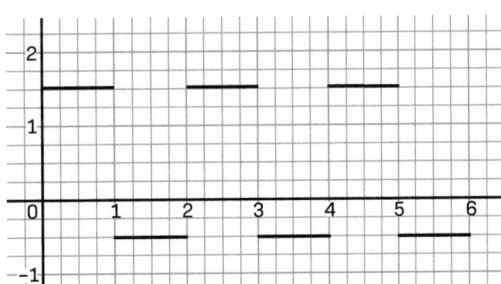

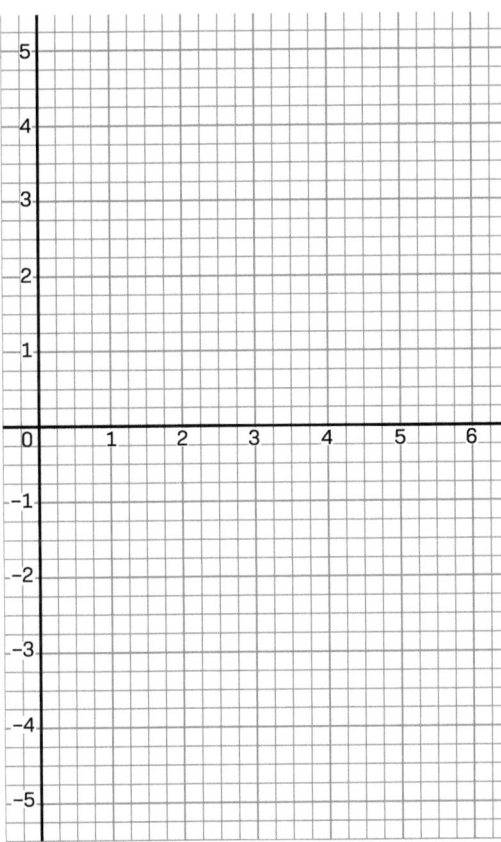

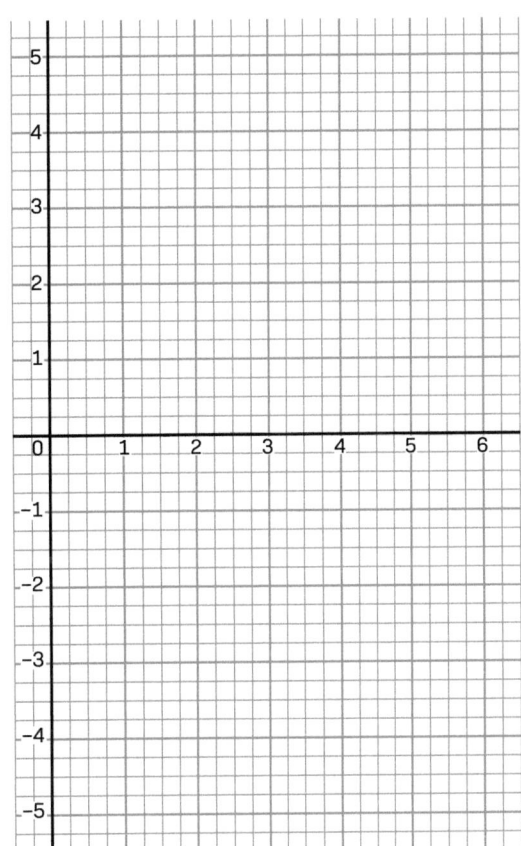

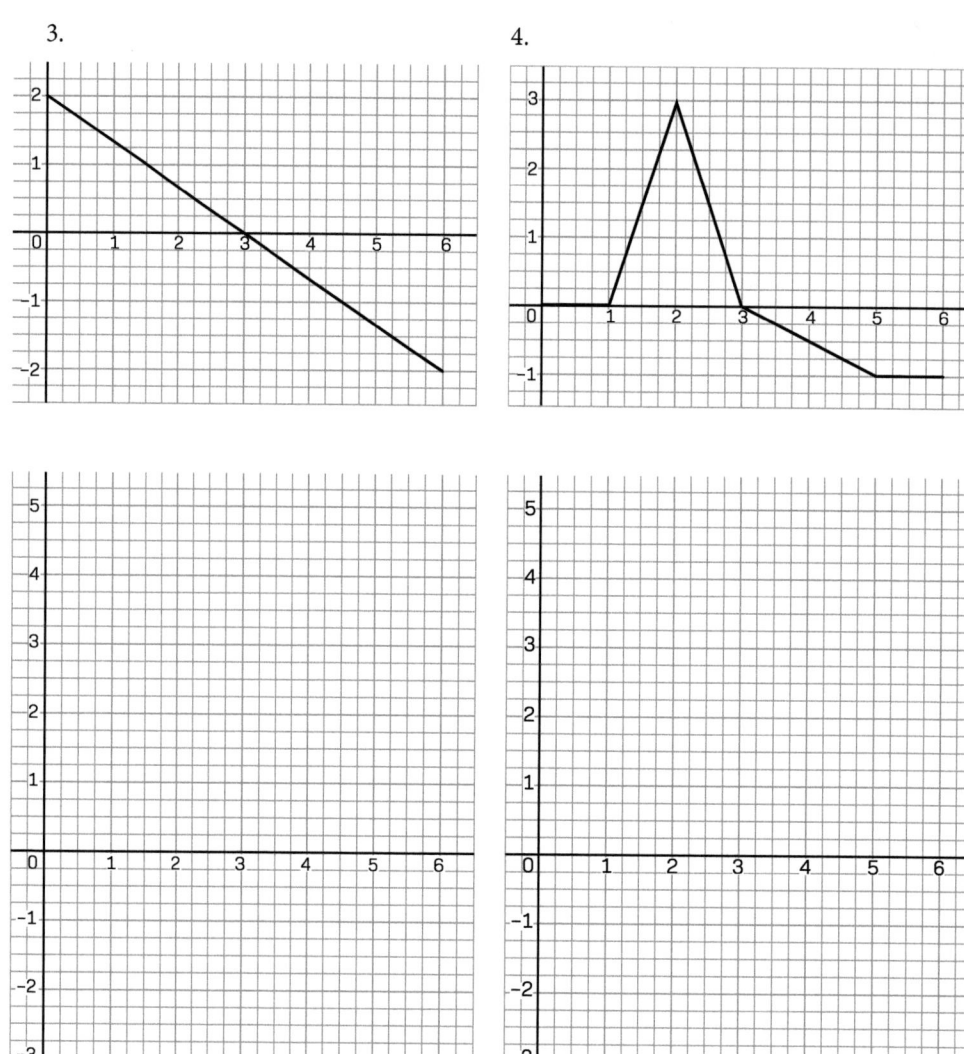

5.

6.

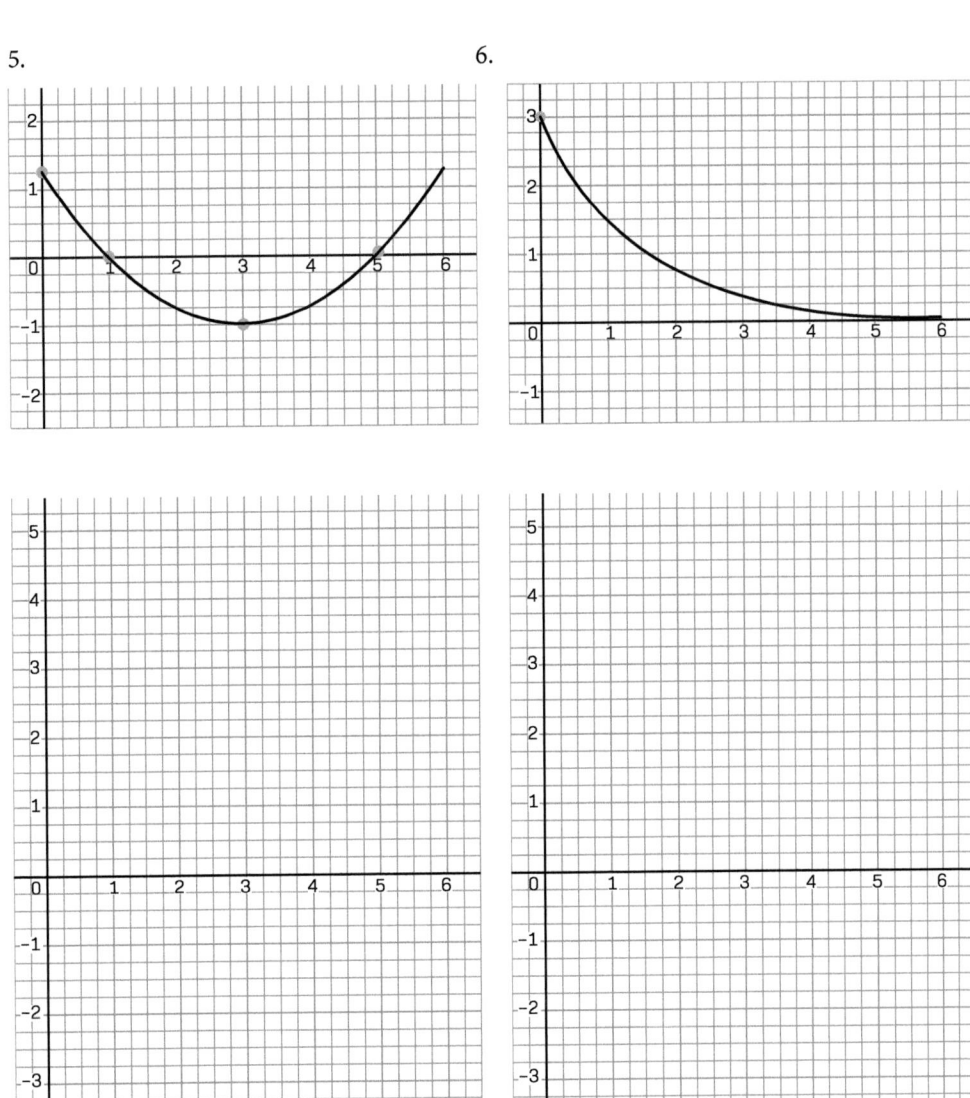

4.5 The Fundamental Theorem of Calculus

OBJECTIVES FOR SECTION 4.5: Upon completing this section, you will be able to do the following:

- Understand the relationship between antiderivatives, indefinite integrals, and derivatives
- Find a formula for an antiderivative of a function that passes through a specified point
- Use the Fundamental Theorem of Calculus to determine the exact value of a definite integral.

In this section, we will discuss and apply the Fundamental Theorem of Calculus, or FTC. This theorem highlights the relationship between the two branches of calculus: differential calculus and integral calculus.

Use the GeoGebra applet "4.5 Fundamental Theorem of Calculus, Part 1" at https://ggbm.at/BN8D2duN to investigate part 1 of the FTC. This file comes preset with the function $f(t) = 2t$, sliders a and b, and a calculation of the definite integral $\int_a^b f(t)dt$. In figure 4.5, we see that the value of this definite integral is 3 when $a = 1$ and $b = 2$.

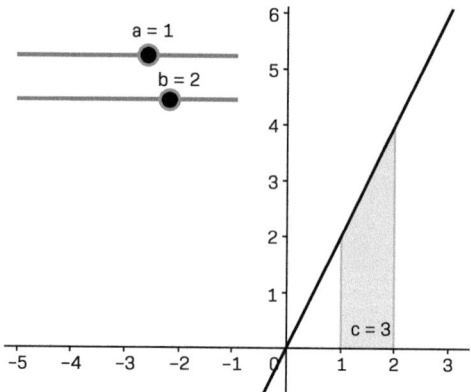

FIGURE 4.5. Using GeoGebra to investigate part 1 of the FTC.

Exercise 4.22: The point B has been constructed to have a horizontal coordinate equal to the value of b, and a vertical coordinate equal to the value of the definite integral. Click the circle next to point B in the Algebra window to view point B. As you move the slider for b, notice how point B moves.

To view the path (or locus) of point B as b varies, click on the circle by "antiderivative" in the Algebra window. You can move the slider for b to note that point B does indeed trace this path. As the name suggests, this function is an antiderivative of $f(t)$. More specifically, we can name this antiderivative F, and state that $F(b) = \int_1^b f(t)dt$.

What happens when you move the slider for *a*?

Notice that doing so shifts the antiderivative vertically. This supports the finding in the last section that the function *f(t)* has many antiderivatives, and these antiderivatives are all related to each other in that their formulas only differ by a constant. Therefore, each of these many antiderivatives all have the same derivative.

The Fundamental Theorem of Calculus, Part 1

If *f(x)* is a function that is continuous on $a \leq x \leq b$, then the function $F(x) = \int_a^x f(t)\,dt$ is continuous on $a \leq x \leq b$, differentiable on $a < x < b$, and $F'(x) = f(x)$.

In other words, *f* is the derivative of *F*, and *F* is an antiderivative of *f*.

We can put part 1 of the FTC into practice in a few ways. First, suppose we have a function $F(x) = \int_a^x 2t\,dt$. We can use the FTC to find the derivative of *F*. Specifically, this construction[2] means that *F* is an antiderivative of $f(x) = 2x$, and therefore $F'(x) = f(x) = 2x$.

Exercise 4.23: Modify the GeoGebra applet to investigate the function $f(t) = -0.5$.

How would you describe the family of antiderivatives you obtain?

If $F(x) = \int_a^x -0.5\,dt$, what is $F'(x)$?

Exercise 4.24: Modify the GeoGebra applet to investigate the function $f(t) = 2 - t^2$.

How would you describe the family of antiderivatives you obtain?

If $F(x) = \int_a^x (2 - t^2)\,dt$, what is $F'(x)$?

---

[2] This function may look strange to us, because the input *x* is the upper limit of integration. We do this because the output (the value of the definite integral) depends on the input, which is the value of *b*. For this reason, the variable for the definite integral needs to be different; I use *t* in these examples to show that the input to obtain *y*-values for the function is different than the input for determining the value of the definite integral. Later, to make connections, I switch from *t* to *x* to describe the function *f*. This is allowable, because if $f(t) = 2t$, then $f(x) = 2x$.

Exercise 4.25: Modify the GeoGebra applet to investigate the function $f(t) = e^t$.

How would you describe the family of antiderivatives you obtain?

If $F(x) = \int_a^x e^t \, dt$, what is $F'(x)$?

Part 1 of the FTC states that the derivative of the antiderivative of a function is that function. In this way, the antiderivative and the derivative "undo" each other—they are like inverse operations. In fact, this is one of the key relationships of the two branches of calculus. In a sense, the definite integral and the derivative are inverses of each other.

In chapters 3 and 4, we have discussed definite integrals. In this section, we can use the idea of a family of antiderivatives to find formulas for the *indefinite integral* of a function. Using the function $f(t) = -0.5$, each antiderivative appeared to be a line. Each line had a slope of -0.5 and a different y-intercept. Therefore, we could write a formula for the antiderivative of $f(t) = -0.5$: $F(x) = -0.5x + C$, where C is a real number. Notice that, regardless of the value of C, $F'(x) = -0.5$, and therefore $F'(x) = f(x)$.

In the initial example with $f(t) = 2t$, the antiderivatives appeared to be parabolas. In fact, each antiderivative has a formula $F(x) = x^2 + C$, where C is a real number.

We use the integral sign without upper and lower limits to denote an indefinite integral:

$$\int -0.5 \, dx = -0.5x + C \text{ and } \int 2x \, dx = x^2 + C$$

We can use our knowledge of derivatives of certain types of functions to help create a list of antiderivatives of functions. For example, if the function is $f(x) = x^3$, then the derivative is $f'(x) = 3x^3$. We can turn this around and state that the antiderivatives of $f(x) = 3x^2$ look like $F(x) = x^3 + C$. Using indefinite integrals, this is stated as $\int 3x^2 \, dx = x^3 + C$.

What if we wanted to find $\int x^2 \, dx$? Notice that x^2 is one-third of $3x^2$, so an antiderivative of x^2 should be $\frac{1}{3} \int 3x^2 \, dx = \frac{1}{3}x^3 + C$.

> **Box 4.8: Teaching Tips**
>
> Help your students deepen their understanding of a topic by regularly returning to it, but provide them with a new twist or application. Remember the list of derivatives that we began in section 4.1? Recreate it here, and use it to determine antiderivatives.

Exercise 4.26: Use table 4.1 to record the derivative of each of the functions. Use table 4.2 to switch things around and record an antiderivative for each of the functions. (Assume a is positive and b is real.)

TABLE 4.1 Derivative Formulas

Function	Derivative
1	
x	
bx	
x^2	
$\frac{1}{2}x^2$	
x^3	
$\frac{1}{3}x^3$	
x^4	
$\frac{1}{4}x^4$	
x^n	
$\sqrt{x}$	
a^x	
$\frac{1}{\ln(a)}a^x$	
e^x	
$\ln(x)$	

TABLE 4.2 Antiderivative Formulas

Function	Antiderivative
1	
b	
x	
x^2	
x^3	
x^4	
x^n	
$\sqrt{x}$	
a^x	
e^x	
$\frac{1}{x}$ (assuming $x > 0$)	

The theorems introduced in section 3.5 apply to indefinite integrals as well. Specifically, if we wish to find antiderivatives of the sum of functions, we can find the sum of the antiderivatives.

$$\int (2x - 0.5)dx = \int 2x\,dx - \int 0.5\,dx = x^2 - 0.5x + C$$

Notice that I simply wrote a single *C* at the end for the constant. This is because the sum of two constants is still a constant, and *C* plays a "generic" constant role. If we know the value of an antiderivative at a particular input, we may be able to find a single formula for *the* antiderivative that passes through that specific point. For example, let $k(x) = 2x - 0.5$. We can find an antiderivative $K(x)$ (that is, $K'(x) = k(x)$) such that $K(2) = 5$.

From the work shown, we know that antiderivatives of $k(x)$ look like $K(x) = x^2 - 0.5x + C$, where C is a constant.

In this case, $K(2) = 2^2 - 0.5(2) + C = 4 - 1 + C = 3 + C$.

We combine this with the given information that $K(2) = 5$ to obtain $5 = 3 + C$, so that $C = 2$.

Therefore, the formula for the antiderivative we seek is $K(x) = x^2 - 0.5x + 2$.

We may check our answer by verifying that $K'(x) = k(x)$ and $K(2) = 5$.

Now, let's find an antiderivative of $p(x) = x^2 + \frac{1}{x} + e$, called $P(x)$, so that $P(1) = e$.

$$\int \left(x^2 + \frac{1}{x} + e\right) dx = \frac{1}{3}x^3 + \ln(x) + ex + C$$

If $P(x) = \frac{1}{3}x^3 + \ln(x) + ex + C$, then

$$P(1) = \frac{1}{3}(1)^3 + \ln(1) + e \cdot 1 + C = \frac{1}{3} + 0 + e + C = \frac{1}{3} + e + C.$$

Combine this with $P(1) = e$ to obtain the equation $e = \frac{1}{3} + e + C$.

When we solve this equation for C, we get $C = -\frac{1}{3}$.

Therefore, the antiderivative we seek is $P(x) = \frac{1}{3}x^3 + \ln(x) + ex - \frac{1}{3}$.

In section 4.4, we saw that we could write the net change from the definite integral in terms of the difference of two values of the antiderivative. We began with $F(b) = F(a) + \int_a^b f(t) dt$.

Using algebra, we can isolate the definite integral to obtain $\int_a^b f(t) dt = F(b) - F(a)$. This is known as part 2 of the FTC.

The Fundamental Theorem of Calculus, Part 2

If $f(x)$ is a function that is continuous on $a \leq x \leq b$, and F is an antiderivative of f, then $\int_a^b f(x) dx = F(b) - F(a)$.

The notation $F(x)\big|_a^b$ is often used when evaluating definite integrals. $F(x)\big|_a^b$ is a compact way to express $F(b) - F(a)$.

Box 4.9: Possible Pitfalls

For example, $x^2 \big|_3^4 = 4^2 - 3^2 = 16 - 9 = 7$.

Suppose $F(x) = \sqrt{x^2 + 1}$. We can use the chain rule to verify that $F'(x) = \frac{x}{\sqrt{x^2 + 1}}$.

Furthermore, we can use part 2 of the FTC to evaluate the definite integral $\int_2^3 \frac{x}{\sqrt{x^2+1}} dx$.

The integrand in the definite integral is $F'(x)$, so an antiderivative would be $F(x)$.

$$\int_2^3 \frac{x}{\sqrt{x^2+1}} dx = \sqrt{x^2+1} \Big|_2^3$$
$$= \sqrt{3^2+1} - \sqrt{2^2+1}$$
$$= \sqrt{10} - \sqrt{5}$$

This is the exact value of the definite integral. A calculator (or other technology) shows the approximate value (to three decimal places) is 0.926.

Here's another problem in the same vein. Suppose $G(x) = e^{x^3}$. Verify that $G'(x) = e^{x^3} \cdot 3x^2$, and then find the exact value of $\int_{-2}^2 e^{x^3} \cdot 3x^2 \, dx$.

The integrand in the definite integral is $G'(x)$, so an antiderivative would be $G(x)$.

$$\int_{-2}^2 e^{x^3} \cdot 3x^2 \, dx = e^{x^3} \Big|_{-2}^2$$
$$= e^{2^3} - e^{(-2)^3}$$
$$= e^8 - e^{-8}$$

Exercise 4.27: Find the exact value of these definite integrals. The first has been worked as an example.

$$\int_0^1 (2e^x - 3) dx \qquad \int_{-1}^1 x^5 \, dx \qquad \int_1^2 (\pi^x \ln(\pi) - x^\pi) dx$$

$$\int_0^1 (2e^x - 3) dx = 2e^x - 3x \Big|_0^1$$
$$= \left(2e^1 - 3(1)\right) - \left(2e^0 - 3(0)\right)$$
$$= (2e - 3) - (2 \cdot 1 - 0)$$
$$= 2e - 3 - 2$$
$$= 2e - 5$$

Note that if we had used a different antiderivative of $2e^x - 3$, such as $2e^x - 3x + \pi$, we would still obtain the same value for the definite integral.

$$\int_0^1 (2e^x - 3) dx = 2e^x - 3x + \pi \Big|_0^1 = \left(2e^1 - 3(1) + \pi\right) - \left(2e^0 - 3(0) + \pi\right)$$
$$= (2e - 3 + \pi) - (2 \cdot 1 - 0 + \pi) = 2e - 3 + \pi - 2 - \pi = 2e - 5$$

Therefore, it is usually most convenient to use $C = 0$ when choosing an antiderivative to use with part 2 of the FTC.

Problem Set 4.5

For items 1–3, use part 1 of the FTC to find the derivative of F(x).

1. $F(x) = \int_0^x (1 - t^3) dt$

2. $F(x) = \int_0^x \sqrt{2 + t^2}\, dt$

3. $F(x) = \int_2^x \ln(t^2 + t^{-2}) dt$

Find the indefinite integrals in problems 4–7.

4. $\int 4x\, dx$

5. $\int (5t - 1.3) dt$

6. $\int 2\pi e^x\, dx$

7. $\int (4t + 4^t + t^4) dt$

8. Let $g(x) = -2$. Find an antiderivative $G(x)$ with $G'(x) = g(x)$ such that $G(0) = 10$.
9. Let $j(x) = \sqrt[3]{x}$. Find an antiderivative $J(x)$ with $J'(x) = j(x)$ such that $J(1) = 4$.
10. Let's explore a more complicated function.
    a. Faith says that if $F(x) = (x^2 + 1)^3$, then $F'(x) = 6x(x^2 + 1)^2$. Do you agree with Faith?
    b. Assume that Faith is correct. Find the value of $\int_1^2 6x(x^2 + 1)^2 dx$.
11. Here's another complicated function to explore.
    a. Brian says that if $F(x) = 4^{x^2}$, then $F'(x) = 2x \cdot 4^{x^2} \ln(4)$ Do you agree with Brian?
    b. Assume that Brian is correct. Find the value of $\int_{-1}^{1} (2x \cdot 4^{x^2} \ln(4)) dx$.

Use part 2 of the FTC to evaluate the definite integrals in items 12–17. Give exact answers.

12. $\int_1^6 \frac{1}{x} dx$

13. $\int_{-1}^{4} (2 - 2x) dx$

14. $\int_0^2 (3x^3 - 2x^2 - e^x) dx$

15. $\int_{-1}^{1} 5e^t\, dt$

16. $\int_1^3 2\pi r\, dr$

17. $\int_4^9 \sqrt{x}\, dx$

References

Common Core State Standards Initiative. (n.d.). Retrieved from http://www.corestandards.org

HooplaKidzLab. (2013, December 11). Egg floating in salt water experiment [Video file]. Retrieved from http://youtube.com/watch?v=zszw6uCiQpc

officalpsy. (2012, July 15). PSY-Gangnam style [Video file]. Retrieved from https://www.youtube.com/watch?v=9bZkp7q19f0

YouTube Trends (2012, September 12). Gangnam Style vs. Call Me Maybe: A popularity comparison. Retrieved from http://youtube-trends.blogspot.com/2012/09/gangnam-style-vs-call-me-maybe.html

Credits

Fig. 4.1: Copyright © by Desmos, Inc.
Fig. 4.2: Copyright © by Desmos, Inc.

Limits, Derivatives, and Integrals

Contexts Within and Beyond Middle School

Think

You may have heard that measurements are approximations. What do you think that means? We will uncover this by attempting to measure different attributes of objects—both familiar and novel, such as the area of a circle and the length of a parabolic arc. You may recall formulas to calculate the volumes of prisms, pyramids, cylinders, and spheres. Where do those formulas come from? We'll investigate the role of calculus in giving meaning to the formulas. We'll also try to optimize measurements—such as finding the lowest price for an item or obtaining the largest volume for a container when using a fixed amount of material.

Remember

To prepare for this chapter, you may want to think about the following topics. They serve as prerequisites and primers to the mathematical content in the chapter.

- Recall the formulas for the circumference and area of a circle with a given radius.
- Determine the volume of a rectangular prism, such as a box that has a base that measures 12 inches by 24 inches and has a height of 18 inches.
- Use the Pythagorean theorem to find the length of the third side of a right triangle, given the lengths of the other two sides. For example, if the longest side is 10 cm and the shortest side is 4 cm, how long is the third side?

Connect

In the middle grades, students begin to seriously consider their future vocation. Measurement and optimization are used in diverse career fields, from manufacturing and finance, to agriculture and the fine arts. While calculus provides efficient methods for optimization, there are alternative methods that are accessible to students in the middle grades, such as creating a table of values and searching for patterns. According to the Common Core State

Standards for Mathematics (CCSS-M) (n.d.) and the Texas Essential Knowledge and Skills (TEKS) (n.d.), students in grades 4 through 8 solve real-world and mathematical problems that involve perimeter and area of two-dimensional figures and surface area and volume of three-dimensional figures.

5.1 Area of a Circle

OBJECTIVES FOR SECTION 5.1: Upon completing this section, you will be able to do the following:

- Understand several ways to derive the formula for the area of a circle
- Approximate the value of π to a specified level of precision

We can use ideas from calculus to find the areas of many different kinds of shapes. In this section, we will examine several ways to approach finding the area of a circle.

Watch the video (Jones, 2015) at http://youtu.be/Fl9lRHw9SDE?hd=1. Dr. John Huber created these animations to illustrate that the area of a circle with radius r is between $2r^2$ and $4r^2$. The final part of the video shows a circle divided into sectors. When those sectors are rearranged, they form a shape that is approximately a parallelogram, with a height of r and a base equal to half of the circumference of the circle.

We can introduce students to these concepts using folded paper and fraction circles. We may also apply calculus to the problem of finding the area of a circle, but you may wish to avoid calculus when speaking to elementary school students.

In section 1.4, we discussed Archimedes's work with the circle. We can use his method of using polygons to approximate a circle. We may examine a sequence of areas of inscribed polygons and compare that to the sequence of areas of circumscribed polygons. Both these sequences approach the same limit.

We may also use calculus to show that the area of a circle with radius r is equal to the sum of circumferences of circles with radii that vary from 0 to r. Using integral notation and ideas from problem 1 of problem set 3.4, we have the area of a circle with radius r is found as follows.

$$\begin{aligned}\text{Area of circle with radius } r &= \int_0^r \left(\text{circumference of circle with radius } x \right) dx \\ &= \int_0^r 2\pi x\, dx \\ &= 2\pi \left(\frac{1}{2} x^2 \right) \Big|_0^r \\ &= \pi x^2 \Big|_0^r \\ &= \pi r^2 - \pi \cdot 0^2 \\ &= \pi r^2 \end{aligned}$$

Here, calculus tells us that the area of a circle with radius r is equal to πr^2.

If we center a circle with radius r at the origin of the Cartesian plane, then each point on the circle is equidistant from the origin. That is, for every point (x, y) on the circle, the distance from that point to the origin is r. We may draw a few lines in to create a right triangle, as shown in figure 5.1, which leads to using the Pythagorean theorem to state that $x^2 + y^2 = r^2$ is the equation for the circle.

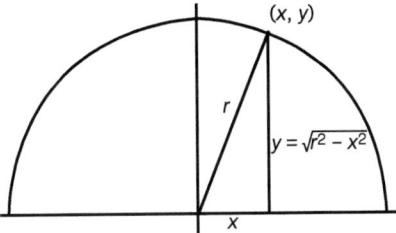

FIGURE 5.1. Plotting a semicircle on the Cartesian plane.

We may solve this equation for y to obtain $y = \pm\sqrt{r^2 - x^2}$. If we use the positive root, we can graph the function $f(x) = \sqrt{r^2 - x^2}$ to make the semicircle shown in figure 5.1. Now that we have a formula for the function, we can use left-hand sums and right-hand sums (as in section 3.2) to approximate the area of the semicircle. The area of the full circle would then be twice the area of the semicircle.

Finally, we could use ideas from the Fundamental Theorem of Calculus. Suppose you know a smart guy named Stewart who says if $f(x) = \frac{1}{2}x\sqrt{r^2 - x^2} + \frac{1}{2}r^2 \sin^{-1}\left(\frac{x}{r}\right)$, then $f'(x) = \sqrt{r^2 - x^2}$. If we believe Stewart, then the area of a quarter circle would be found in this way.

Box 5.1: Teaching Tips

You can introduce ideas to students through hypothetical people. That way, their reasoning can be questioned without hurting anyone's feelings.

$$\int_0^r \sqrt{r^2 - x^2}\, dx = \int_0^r f'(x)\, dx$$

$$= f(x)\Big|_0^r$$

$$= \frac{1}{2}x\sqrt{r^2 - x^2} + \frac{1}{2}r^2 \sin^{-1}\left(\frac{x}{r}\right)\Big|_0^r$$

$$= \left[\frac{1}{2}r\sqrt{r^2 - r^2} + \frac{1}{2}r^2 \sin^{-1}\left(\frac{r}{r}\right)\right] - \left[\frac{1}{2}(0)\sqrt{r^2 - 0^2} + \frac{1}{2}r^2 \sin^{-1}\left(\frac{0}{r}\right)\right]$$

$$= \left[\frac{1}{2}r\sqrt{0} + \frac{1}{2}r^2 \sin^{-1}(1)\right] - \left[\frac{1}{2}(0)\sqrt{r^2} + \frac{1}{2}r^2 \sin^{-1}(0)\right]$$

$$= \left[0 + \frac{1}{2}r^2 \cdot \frac{\pi}{2}\right] - \left[0 + \frac{1}{2}r^2 \cdot 0\right] \quad \text{when using radians with } \sin^{-1}$$

$$= \frac{1}{4}\pi r^2 - 0$$

$$= \frac{1}{4}\pi r^2$$

Therefore, the area of a full circle would be four times the area of the quarter circle, or πr^2.

Problem Set 5.1

For items 1–3, consider the "Area of a Circle" video referenced at the beginning of this section.

1. How do the animations show that the area of a circle with radius r is less than $4r^2$?
2. How do the animations show that the area of a circle with radius r is greater than $2r^2$?
3. In the final animation, we see that the circle with radius r and circumference C can be transformed into a figure that resembles a parallelogram. The area for the parallelogram would be $A = \frac{1}{2} C \cdot r$. This formula for area is written in terms of C and r. Replace C with a formula that involves r to obtain a formula for area in terms of r alone.
4. Use the GeoGebra applet "5.1 Area of a Circle" at https://ggbm.at/BN8D2duN to investigate the area of a circle, following Archimedes's method of doubling the number of sides. Note that this file has a circle centered at the origin with a **radius of 1 unit**. Regular polygons with n sides have been inscribed in and circumscribed about the circle.

    a. Complete the following table:

Number of sides, n	Area of inscribed polygon	Area of circumscribed polygon
6		
12		
24		
48		
96		
192		
384		
Limit as n → ∞		

    b. Using $n = 400$, this GeoGebra file allows us to approximate the value of π as _____, which is correct to _____ decimal places.

5. Use algebra to show how we can transform the equation of a circle $x^2 + y^2 = r^2$ into the function of a semicircle $f(x) = \sqrt{r^2 - x^2}$.
6. Use GeoGebra to calculate left-hand sums and right-hand sums.

    a. Complete the table of values of left-hand sums and right-hand sums to estimate the value of the definite integral $\int_{-1}^{1} \sqrt{1 - x^2}\, dx$. Write at least 4 decimal places.

NUMBER OF SUBDIVISIONS	2	10	50	100	500
LEFT-HAND SUM					
RIGHT-HAND SUM					

    b. The area of a circle with radius 1 is π, but these sums do not approach π. Why not?

5.2 Extrema

OBJECTIVES FOR SECTION 5.2: Upon completing this section, you will be able to do the following:

- Identify absolute extrema, relative extrema, and inflection points on the graph of a function
- Find the critical values of a function
- Determine the location of absolute extrema, relative extrema, and inflection points of a function

In this section, we discuss features of a function, such as where the minimum or maximum values occur. These extreme values are collectively referred to as *extrema*, which is a plural word. There are two types of extrema that we consider: *absolute extrema* (including *absolute minimum* and *absolute maximum*) and *relative extrema* (including *relative minimum* and *relative maximum*). The plural form of minimum and maximum are *minima* and *maxima*, respectively.

Absolute extrema

A function $f(x)$ may have an *absolute maximum* value or an *absolute minimum* value.

f has an absolute maximum if there is an M in the domain of f such that

$f(M) \geq f(x)$ for all x in the domain of f.

f has an absolute minimum if there is an m in the domain of f such that

$f(m) \leq f(x)$ for all x in the domain of f.

Exercise 5.1:

a. Do either of the functions shown have an absolute maximum?
b. If so, estimate the values of M and $f(M)$. If not, explain why not.
c. Do either of the functions shown below have an absolute minimum?
d. If so, estimate the values of m and $f(m)$. If not, explain why not.

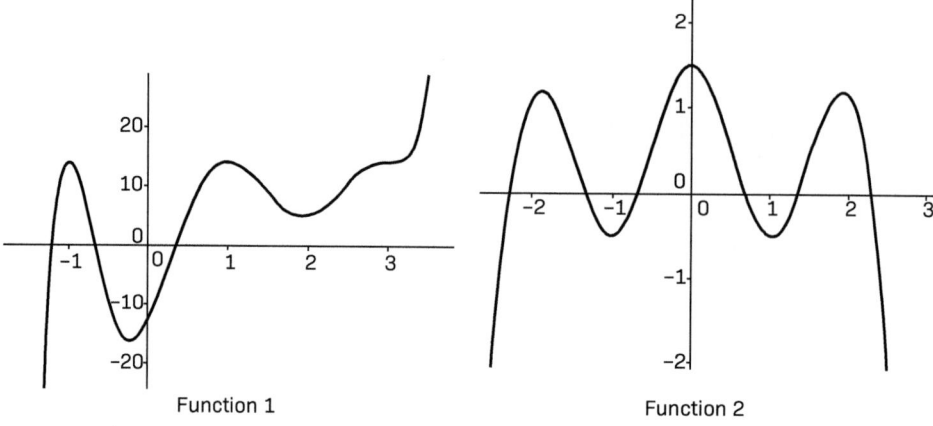

Function 1   Function 2

FIGURE 5.2. Identifying extrema in functions.

Box 5.2: Possible Pitfalls

Assume that the only turns in the graphs of the functions in figure 5.2 are shown. There is nothing interesting hiding beyond the displayed portion.

Exercise 5.2: Let's restrict the domains of these functions to $-0.5 \leq x \leq 2$. Complete the charts that follow for each function shown.

Function 1

Value	Estimate
M	
f(M)	
m	
f(m)	

Function 2

Value	Estimate
M	
f(M)	
m	
f(m)	

You should notice that the function on the left appears to have no absolute minimum or maximum if we consider the domain to be all real numbers. The function on the right does appear to have an absolute maximum at $M = 0$, where $f(0) = 1.5$. When we restrict the domains of the functions, we obtain absolute minima and absolute maxima in both cases.

Relative extrema

Consider a function $f(x)$ and a number c in the domain of f.
Informally, the value $f(c)$ is a *relative maximum* if $f(c)$ is greater than the values of f for x-values near c.
Similarly, $f(c)$ is a *relative minimum* if $f(c)$ is less than the values of f for x-values near c.
Relative maxima and minima are collectively referred to as *relative extrema*.

Exercise 5.3: Identify the relative maxima and relative minima in the graph of the following function and estimate the values. You may not need all of the c's.

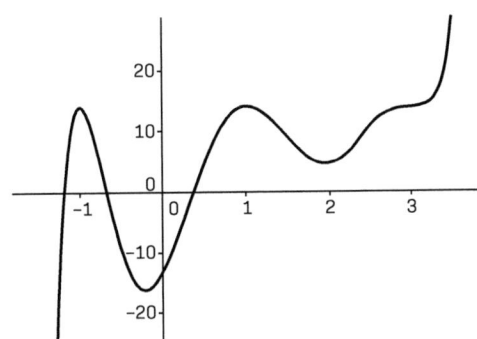

Value	Estimate	Value	Estimate	Value	Estimate
c_1		$f(c_1)$		$f'(c_1)$	
c_2		$f(c_2)$		$f'(c_2)$	
c_3		$f(c_3)$		$f'(c_3)$	
c_4		$f(c_4)$		$f'(c_4)$	
c_5		$f(c_5)$		$f'(c_5)$	
c_6		$f(c_6)$		$f'(c_6)$	

Exercise 5.4: Identify the relative maxima and relative minima in the graph of the following function and estimate the values. You may not need all of the c's.

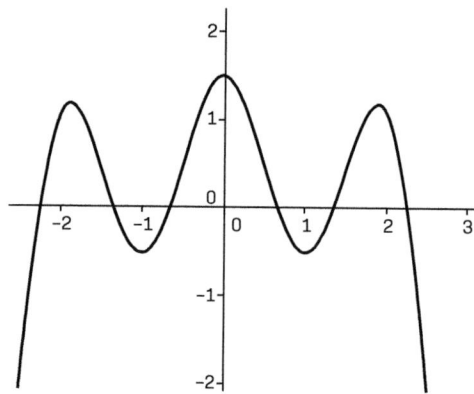

Value	Estimate	Value	Estimate	Value	Estimate
c_1		$f(c_1)$		$f'(c_1)$	
c_2		$f(c_2)$		$f'(c_2)$	
c_3		$f(c_3)$		$f'(c_3)$	
c_4		$f(c_4)$		$f'(c_4)$	
c_5		$f(c_5)$		$f'(c_5)$	
c_6		$f(c_6)$		$f'(c_6)$	

The values of x where $f'(x) = 0$ are called the *critical values* of f.

Exercise 5.5: In exercises 5.3 and 5.4, are all of the values of c that you found critical values? Are there any critical values c in the functions shown where $f(c)$ is not a relative maximum or a relative minimum?

On Function 1 in figure 5.2 and exercise 5.3, note that $f'(3)$ appears to be 0, but $f(3)$ is neither a relative maximum nor a relative minimum. Therefore, we see evidence that the following statements are true:

- If $f(c)$ is a relative maximum or relative minimum of f, then $f'(c) = 0$, and c is a critical value of f.
- If $f'(c) = 0$, then c is a critical value of f, but $f(c)$ is not necessarily a relative maximum or minimum of f. You would need to do some further checking to see if $f(c)$ is a relative maximum, relative minimum, or neither.

Another type of interesting point occurs when the function changes concavity. These are called *inflection points*. If the point $(c, f(c))$ is an inflection point, then $f''(c) = 0$. On the other hand, if $f''(c) = 0$, you need to check to see if the point $(c, f(c))$ is actually an inflection point.

Let's apply these definitions to a specific function. Consider $f(x) = 2x^2 + 12x - 10$.

a. Find all of the critical values of the function.
b. Does this function have any relative extrema? If so, identify them using coordinates.
c. Does this function have any inflection points? If so, identify them using coordinates.

Here are my solutions to the tasks in parts a, b, and c.

a. The number c is a critical value of the function f if $f'(c) = 0$. Therefore, we first find the derivative of f.
   $f'(x) = 4x + 12$, so $f'(c) = 0$ implies that $4c + 12 = 0$. This leads to $c = -3$. Therefore, this function has only one critical value.
b. We can tell from the formula of the function (or by looking at the graph), that the graph of this function is a parabola that is concave up. Therefore, the critical value of -3 corresponds to the relative minimum at $(-3, f(-3)) = (-3, -28)$.
c. If there were any inflection points, they would occur where the second derivative is equal to zero. Note that $f''(x) = 4$, and will therefore never be zero. That means that this function has no inflection points.

Exercise 5.6: Repeat this process (parts a, b, and c) using the following functions:

$g(x) = -3x + 2$ $\qquad$ $h(x) = x^4 - 8x^3 + 16x^2 - 8$

$j(x) = 0.25x^2 + 2\sqrt{x}$ $\qquad$ $k(x) = e^x - x$

We can apply this to real-world phenomena, such as projectile motion. If an object is thrown from a height h_0 with an initial velocity v_0, then the height of the object is a function of time. If the height is measured in feet, time in seconds, and velocity in feet per second, then the height can be found using $f(t) = h_0 + v_0 t - 16t^2$. The object will reach its maximum height when $f'(t) = 0$. The derivative $f'(t) = v_0 - 32t$ is equal to 0 when $t = \dfrac{v_0}{32}$.

Problem Set 5.2

Consider the following functions on the given domains. In problems 1–8, identify the following:

   a. The value M in the domain that gives the absolute maximum
   b. The absolute maximum $f(M)$
   c. The value m in the domain that gives the absolute minimum
   d. The absolute minimum $f(m)$

1. $f(x) = 4 + 3x, \ 0 \le x \le 4$
2. $f(x) = 4 + 3x, \ -3 \le x \le 1.5$
3. $f(x) = 4 - x^2, \ -1 \le x \le 2$
4. $f(x) = 4 - x^2, \ 1 \le x \le 5$

5. $f(x) = e^{-x} + x, -1 \leq x \leq 2$
6. $f(x) = \cos(\pi x), -1 \leq x \leq 5$
7. $f(x) = 4x - \ln(x), 1 \leq x \leq 3$
8. $f(x) = 5 + 3x - x^3, -3 \leq x \leq 0$

For each of the functions in items 9–14,
   a. identify all the critical values,
   b. give the coordinates of any relative minima, and
   c. give the coordinates of any relative maxima

9. $f(x) = 4 + 3x$
10. $f(x) = 4 - x^2$
11. $f(x) = e^{-x} + x$
12. $f(x) = 4x - \ln(x)$
13. $f(x) = 5 + 3x - x^3$
14. $f(x) = 6\sqrt{x} - x$
15. Examine the graphs of the functions listed in items 9–14.
    a. Which of these functions appear to have inflection points?
    b. Give the coordinates of the inflection points of these functions.
16. The standard form of a quadratic function is $f(x) = ax^2 + bx + c$, where a, b, and c are real numbers, and $a \neq 0$.
    a. Find the first and second derivatives of this function.
    b. Find the critical values of this function.
    c. Use the second derivative to explain why quadratic functions have no inflection points.
17. The standard form of an exponential function is $f(x) = a \cdot b^x$, where a and b are real numbers, $a \neq 0$, b is positive, and $b \neq 1$.
    a. Find the first and second derivatives of this function.
    b. Use the first derivative to explain why exponential functions have no relative extrema.
    c. Use the second derivative to explain why exponential functions have no inflection points.

5.3 Optimization

OBJECTIVES FOR SECTION 5.3: Upon completing this section, you will be able to do the following:

- Create mathematical formulas to model situations
- Determine a quantity to maximize or minimize
- Apply calculus to find maximum or minimum values

We can use the ideas of extrema from the last section in many real-world applications. The next few pages have some guided investigations about certain topics.

> **Box 5.3: Teaching Tips**
>
> Attract your students' attention by relating mathematics to other things that interest them.

Exercise 5.7: Before we begin the investigations, let's brainstorm about some questions that involve optimization. Read each situation and then think of questions that involve maximizing or minimizing some quantity.

- An artist with a fixed amount of ribbon wants to make a rectangle.

    What is the quantity to optimize?
    Are we interested in the maximum, minimum, or both?

- A farmer with a fixed amount of fencing material wants to make a pen for rabbits.

    What is the quantity to optimize?
    Are we interested in the maximum, minimum, or both?

- A manufacturer is making cylindrical water bottles that hold 1 liter of water.

    What is the quantity to optimize?
    Are we interested in the maximum, minimum, or both?

- A pet store owner needs to provide a guinea pig with a rectangular pen with a fixed amount of floor area.

    What is the quantity to optimize?
    Are we interested in the maximum, minimum, or both?

- A traveler needs to get from one place to another.

    What is the quantity to optimize?
    Are we interested in the maximum, minimum, or both?

On the next few pages, follow the guided investigations to answer some optimization questions.

Exercise 5.8: To make a box, four identical squares will be removed from the corners of a rectangle, and then the sides will be folded. Suppose we complete this process with an index card measuring 3 inches by 5 inches.

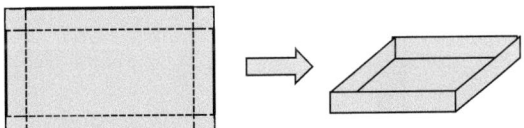

Let x be the length of a side of a square. Complete the table that follows for different values of x.

x	length	width	height	volume
0 in				
0.25 in				
0.5 in				
0.75 in				
1 in				
1.25 in				
1.5 in				
1.75 in				
2 in				

Are any of the values of x in the table not reasonable? Why?
Let's restrict the domain to _____ $\leq x \leq$ _____.
What is the absolute minimum volume on this domain?
Which x values may give the maximum volume of the box?
Write formulas for the length, width, height, and volume as functions of x.

$l(x) =$ $w(x) =$ $h(x) =$

$V(x) =$

Graph $y = V(x)$. Do you see any relative minima or relative maxima?
Write a formula for $V'(x)$.
Solve $V'(x) = 0$ to locate the critical values.
The maximum volume of _____ is obtained when $x =$ _____.

Box 5.4: Possible Pitfalls

If a right triangle is involved, then you may need to apply the Pythagorean theorem. Exercise 5.9 also involves the chain rule from section 4.3.

Exercise 5.9: A farmer has a straight section of fence material that is 4 meters long.
She will position it in the corner of a barn to create a triangular pen.
The walls of the corner (conveniently) form a right angle.
Draw a picture of the triangular pen. The barn walls have been drawn for you.

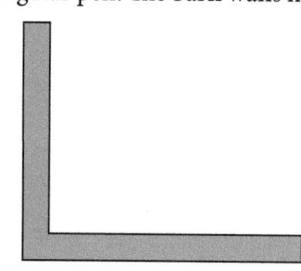

Label the two shorter sides of the triangle as x and y.

Write an equation that shows the relationship between x and y.
Write a formula that gives the area of the triangle as a function of the length x.

$$A(x) =$$

Explain why this formula also gives the area of the triangle: $f(x) = \frac{1}{2}(16x^2 - x^4)^{\frac{1}{2}}$

Verify that $f'(x) = \frac{8x - x^3}{\sqrt{16x^2 - x^4}}$.

What are the critical values of f?
What is the maximum area of the pen? What is the minimum area?

Problem Set 5.3

1. A rectangular piece of paper that measures 8 inches by 11 inches will be formed into a box. Four identical squares will be removed from the corners of a rectangle, and then the sides will be folded. What is the maximum volume of the box?
2. A farmer will build a rectangular pen with 64 meters of fencing material. What is the maximum area that she can enclose? What are the dimensions (length and width) that give that maximum area?
3. A gardener wishes to make a rectangular garden with an area of 20 square feet. What dimensions (length and width) minimize the perimeter of the rectangle? What is the minimum perimeter?
4. A square of paper, measuring 4 inches on each side, has isosceles triangles removed from it. The remaining paper is folded into a square pyramid. In the diagrams that follow, x represents the height of the removed triangles.

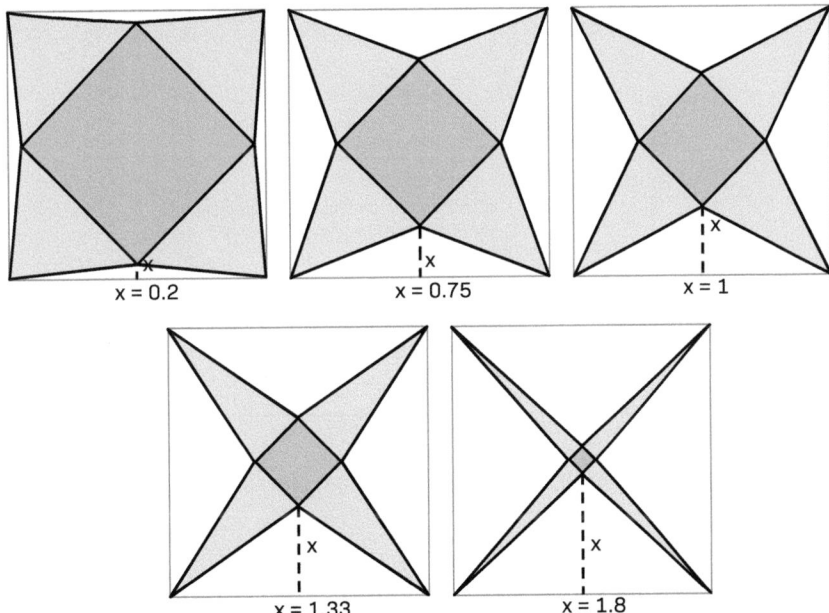

The volume, V, of the resulting square pyramid is a function of x, given by the formula

$$V(x) = \frac{2}{3}\left(8\sqrt{x} - 8x^{1.5} + 2x^{2.5}\right).$$

a. Find the derivative of V with respect to x.
b. An equivalent formula (possibly different from the one you obtained in part a.) for the derivative of V with respect to x is $\dfrac{dV}{dx} = \dfrac{2\left(4 - 12x + 5x^2\right)}{3\sqrt{x}}$.
Use this derivative to find the values of x that minimize and/or maximize the volume of the square pyramid.
c. What are the minimum and maximum volumes of the square pyramid?

5.4 Length of a Curve

OBJECTIVES FOR SECTION 5.4: Upon completing this section, you will be able to do the following:

- Understand the formula for the length of a curve
- Approximate the length of a curve to a specified level of precision

We have seen how calculus can help us find areas of shapes with irregular borders. You may not be surprised to hear that we can also use tools from calculus to help us find perimeters and lengths of curves.

Suppose we have the curve $f(x) = x^2$ on the interval $0 \leq x \leq 2$, and we wish to find the length of this curve on the coordinate plane. This example is illustrated in the GeoGebra applet "5.4 Length of a Curve" at https://ggbm.at/BN8D2duN and shown in figure 5.3.

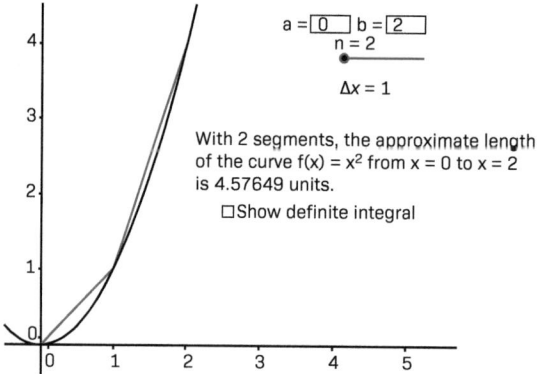

FIGURE 5.3. Using GeoGebra to approximate the length of a curve.

One method to find the length of this curve involves dividing the domain into n equal subintervals, each with a width of Δx. The points on the function corresponding to the endpoints of these subintervals would then be connected with straight line segments. The lengths of these segments are then calculated. The sum of these lengths is used as an approximation of the length of the curve.

As usual, we can improve our estimate by increasing the number of subintervals, which in turn decreases the width of each interval.

Exercise 5.10: Use the GeoGebra applet "5.4 Length of a Curve" to complete the following table.

Number of subintervals n	Width of each subinterval Δx	Sum of lengths of segments
1		
2		
4		
10		
15		
20		

The approximate length of the curve is _____, correct to _____ decimal places.

At this point, it may be helpful to consider this question: How do we find the length of a straight line segment? The answer is found in the distance formula, which is an application of the Pythagorean theorem. Consider the line segment joining the points $A = (x, f(x))$ and $B = (x + \Delta x, f(x + \Delta x))$, as shown in figure 5.4. If we draw in horizontal and vertical lines, a right triangle appears.

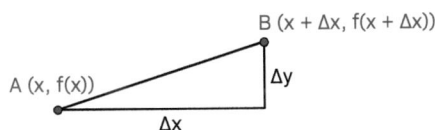

FIGURE 5.4. Using the Pythagorean theorem.

According to the Pythagorean theorem, the length of segment AB may be found using $(AB)^2 = (\Delta x)^2 + (\Delta y)^2$, or $AB = \sqrt{(\Delta x)^2 + (\Delta y)^2}$. At first glance, this may appear complicated. To make it simpler, I will first make it more complicated. Let's multiply the square root by 1, but written in a fancy form: $\dfrac{\sqrt{(\Delta x)^2}}{\sqrt{(\Delta x)^2}}$.

Box 5.5: Teaching Tips

Have your students provide mathematical reasons for the steps they use when solving a problem. This will help you see what they understand, and it will help them be explicit in their process.

Exercise 5.11: Justify each of the steps in this chain of equations.

$$AB = \sqrt{(\Delta x)^2 + (\Delta y)^2} \cdot \frac{\sqrt{(\Delta x)^2}}{\sqrt{(\Delta x)^2}}$$

$$= \frac{\sqrt{(\Delta x)^2 + (\Delta y)^2}}{\sqrt{(\Delta x)^2}} \cdot \sqrt{(\Delta x)^2}$$

$$= \sqrt{\frac{(\Delta x)^2 + (\Delta y)^2}{(\Delta x)^2}} \cdot \Delta x$$

$$= \sqrt{\frac{(\Delta x)^2}{(\Delta x)^2} + \frac{(\Delta y)^2}{(\Delta x)^2}} \cdot \Delta x$$

$$= \sqrt{1 + \left(\frac{\Delta y}{\Delta x}\right)^2} \cdot \Delta x$$

Box 5.6: Possible Pitfalls

Use the GeoGebra applet to calculate the value of the Riemann sums.

To calculate the sum of these lengths, we can use a Riemann sum, similar to $\sum_{i=1}^{n} \left(\sqrt{1 + \left(\frac{\Delta y_i}{\Delta x}\right)^2} \cdot \Delta x \right)$. As n gets larger, Δx gets smaller, and $\frac{\Delta y}{\Delta x}$ approaches $f'(x)$. The derivative appears in the calculation of the length of the curve! The limit of this Riemann sum as n approaches infinity can be represented using a definite integral.

Length of a curve

For the continuous, differentiable function $f(x)$ on the interval $a \leq x \leq b$, the length of the curve is $\int_a^b \sqrt{1 + \left(f'(x)\right)^2}\, dx$.

Problem Set 5.4

1. Use the GeoGebra applet "5.4 Length of a Curve" to complete the following table and approximate the length of the given curve.

   $f(x) = e^x$ on the interval $0 \leq x \leq 2$

Number of subintervals, n	Sum of lengths of segments
1	
2	
5	
10	
15	
20	

   The approximate length of the curve is _____, correct to _____ decimal places.

2. Use the GeoGebra applet "5.4 Length of a Curve" to complete the table that follows and approximate the length of the given curve.

   $f(x) = \sin(10x)$ on the interval $1 \leq x \leq 4$

Number of subintervals, n	Sum of lengths of segments
1	
2	
5	
10	
15	
20	

   The approximate length of the curve is _____, correct to _____ decimal places.

3. Consider the function $f(x) = \ln(x)$ on the interval $1 \leq x \leq 4$.
   a. Write a definite integral to represent the length of this curve.
   b. Use technology to evaluate the integral and therefore find the length of this curve.

4. Consider the function $f(x) = x^2 - 2x + 3$ on the interval $0 \leq x \leq 2$.
   a. Write a definite integral to represent the length of this curve.
   b. Use technology to evaluate the integral and therefore find the length of this curve.

5. Consider the function $f(x) = x^k$ on the interval $0 \leq x \leq 1$.
   Complete the following table, using technology.

EXPONENT K	1	2	3	5	10	20	Limit as $k \to \infty$
LENGTH OF CURVE							

5.5 Solids of Revolution: Volume and Surface Area

OBJECTIVES FOR SECTION 5.5: Upon completing this section, you will be able to do the following:

- Identify solids of revolution
- Understand the formulas for the volume and surface area of a solid of revolution
- Determine the volume and surface area of a solid of revolution

5.5.1 Solids of Revolution

Just as we have learned to use calculus to measure areas and lengths of two-dimensional figures, we can also use similar tools to find volumes and surface areas of three-dimensional figures. We will focus solely on solids that are formed by revolving a region about the x-axis, although some techniques, particularly those for finding volume, could be applied to solids that are not formed by revolution (such as a square pyramid).

We will use the 3D Graphics feature of GeoGebra to investigate solids of revolution. By itself, the 3D Graphics view of GeoGebra provides a box and a plane with axes to indicate three dimensions of space. Figure 5.5 shows the result of graphing a linear function over an interval and then rotating around the x-axis. The applet "5.5 Solid of Revolution" at https://ggbm.at/BN8D2duN will allow you to create this and other solids of revolution, allowing you to input a function and interval for the domain, and then build a wire frame of the solid in the 3D Graphics view using sliders.

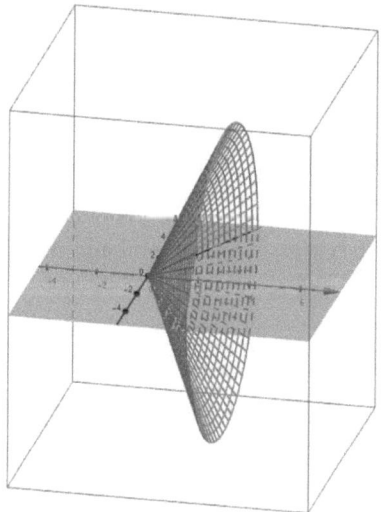

FIGURE 5.5. Creating a solid of revolution.

> **Box 5.7: Teaching Tips**
>
> Focus your students' attention on specific information by giving them options, like a word bank.

Exercise 5.12: For each of the functions that follow, identify the solid (from the word bank) that would be generated by rotating the function about the x-axis. Also draw a sketch of the solid.

Word Bank

cone   cylinder   frustum of cone   pyramid   sphere

$f(x) = 2x, \ 0 \leq x \leq 3$

$g(x) = \dfrac{x}{3}, \ 4 \leq x \leq 5$

$h(x) = \sqrt{4 - x^2}, \ -2 \leq x \leq 2$

$j(x) = 2, \ 1 \leq x \leq 3$

$p(x) = x - 4, \ 4 \leq x \leq 5$

$q(x) = 2 - .1x, \ 0 \leq x \leq 3$

In the previous exercise, you should notice that there are two cones, one cylinder, one sphere, and no pyramids. There are also two solids that look like what remains after removing a small cone from a larger one. This solid is called a frustum of a cone.

5.5.2 Volume of a Solid of Revolution

Now that we have an idea of what the shape appears to be, we move on to finding the volume of the solid. What is the volume of the solid formed by rotating the following function about the x-axis?

$$f(x) = 2x, \ 0 \leq x \leq 3$$

In figure 5.6 (at left) is a graph of the function. If I rotate the triangular region (center picture) between the function and the x-axis about the x-axis, I obtain the cone (shown on the right).

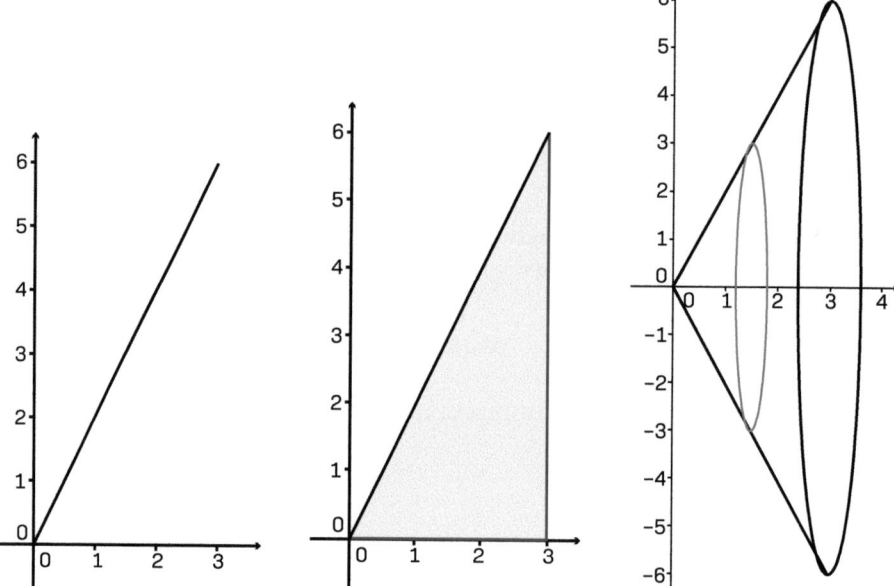

FIGURE 5.6. Rotating a function about the x-axis.

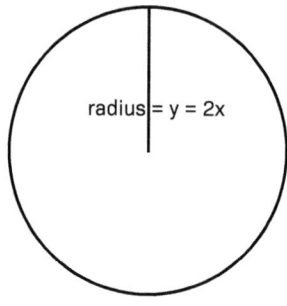

FIGURE 5.7. A circular cross-section.

Each vertical cross-section of the cone is a circle. The radius of each cross section is the height of the function at that point. That means the radius is the y-value, which is $2x$ in this case. (See figure 5.7.) Therefore, the area of each cross-section is $\pi(2x)^2 = 4\pi x^2$. The thickness of each cross-section disk is Δx, so the volume of each cylindrical "chunk" is $(4\pi x^2)\Delta x$.

A Riemann Sum for the volume of the cone is $\sum (4\pi x^2)\Delta x$. This leads to the definite integral for the volume of the cone, $\int_0^3 4\pi x^2 \, dx$.

From this, we obtain $4\pi \left(\frac{1}{3}x^3\right)\Big|_0^3 = 4\pi\left(\frac{1}{3}(3^3)\right) - 4\pi\left(\frac{1}{3}(0^3)\right) = 36\pi$

You may recall that there is a formula for the volume of a cone: $\frac{1}{3}\pi r^2 h$. This particular cone has a radius $r = 6$ and a height $h = 3$. Substituting these values into the formula also yields 36π.

At this point, we can ask a few questions:

1. Will this method work for other solids?
2. Where does the formula $\frac{1}{3}\pi r^2 h$ come from?
3. How can we obtain formulas for other solids?

The brief answer to (1) is yes, given that we are careful about our meaning of "this method." We may find the volume of a solid by imagining it as a set of "chunks" that look like prisms or cylinders, where the cross-section of the solid forms the base of each "chunk." We can multiply the area of this cross section by a small thickness to obtain a volume. Adding these volumes together will give us an approximation of the volume of the solid. To improve our approximation, we can use more "chunks" which, in turn, causes the thickness of each to be smaller.

As you may have figured out, we could use this method to find the volume of a square pyramid, where each cross section is a square. The "chunks" would be small prisms with square bases. The calculation for volume would be different than the preceding example in that the area of a cross section would not be equal to π times the square of a radius, however.

To answer (2), let's look at a general cone with radius r and height h. This could be formed by rotating a line segment about the x-axis. We could start the line segment at the origin and have it pass through the point (h, r), which would mean it would have a slope of $\frac{r}{h}$. A formula to describe this line segment would be $y = \left(\frac{r}{h}\right)x$, on the interval $0 \leq x \leq h$.

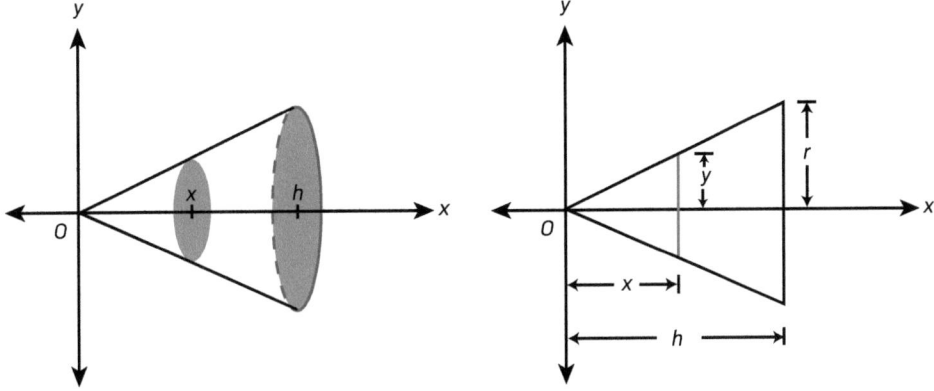

FIGURE 5.8. Finding the volume of a cone.

Each cross-section is a circle, and the radius of each circle is y, which means the area of the cross-section is $\pi y^2 = \pi\left(\left(\frac{r}{h}\right)x\right)^2 = \pi\left(\frac{r^2}{h^2}\right)x^2$. The volume of the solid would then be

$$\int_0^h \pi\left(\frac{r^2}{h^2}\right)x^2\,dx = \pi\left(\frac{r^2}{h^2}\right)\left(\frac{1}{3}x^3\right)\Big|_0^h = \pi\left(\frac{r^2}{h^2}\right)\left(\frac{1}{3}h^3\right) - \pi\left(\frac{r^2}{h^2}\right)\left(\frac{1}{3}(0^3)\right) = \pi(r^2)\left(\frac{1}{3}h\right) = \frac{1}{3}\pi r^2 h.$$

Would you like to find formulas for other solids of revolution, as suggested by question (3)? We can find the volume of a sphere using a similar method. A sphere with radius r can be formed by rotating the semicircle $y = \sqrt{r^2 - x^2}$ on the interval $-r \leq x \leq r$ about the x-axis.

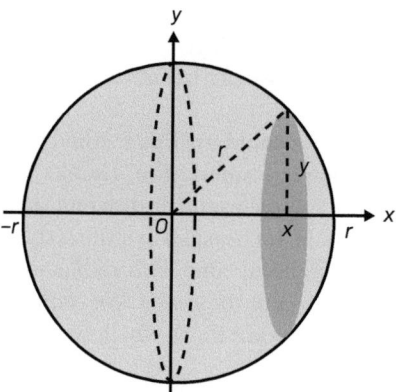

FIGURE 5.9. Finding the volume of a sphere.

Each cross section is a circle, and the radius of each cross section is y. The area of a cross section is $\pi y^2 = \pi\left(\sqrt{r^2 - x^2}\right)^2 = \pi(r^2 - x^2) = \pi r^2 - \pi x^2$. We find the volume with a definite integral.

$$\int_{-r}^{r}\left(\pi r^2 - \pi x^2\right)dx = \left.\left(\pi r^2 x - \pi\left(\frac{1}{3}x^3\right)\right)\right|_{-r}^{r}$$

$$= \left(\pi r^2 r - \pi\left(\frac{1}{3}r^3\right)\right) - \left(\pi r^2(-r) - \pi\left(\frac{1}{3}(-r)^3\right)\right)$$

$$= \left(\pi r^3 - \frac{1}{3}\pi r^3\right) - \left(-\pi r^3 + \frac{1}{3}\pi r^3\right)$$

$$= \pi r^3 - \frac{1}{3}\pi r^3 + \pi r^3 - \frac{1}{3}\pi r^3$$

$$= \left(1 - \frac{1}{3} + 1 - \frac{1}{3}\right)\pi r^3$$

$$= \frac{4}{3}\pi r^3$$

Therefore, the volume of a sphere with radius r is $\frac{4}{3}\pi r^3$. This leads to an interesting extension involving three solids: a cylinder, cone, and hemisphere, each with the same radius and height, as shown in figure 5.10. The "height" of a hemisphere is the radius, so the heights of the cylinder and cone are also equal to the radius. See the problem set for questions related to this set of solids.

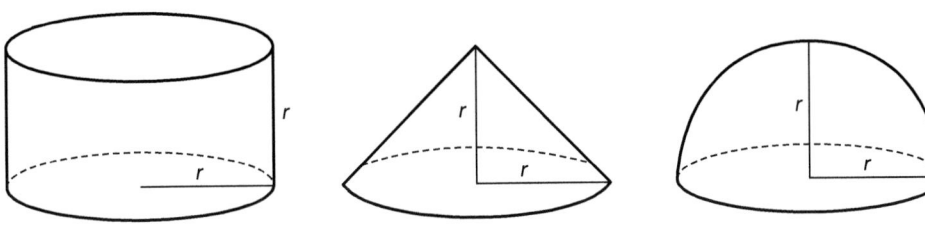

FIGURE 5.10. A related set of solids.

Box 5.8: Possible Pitfalls

This formula *only* works for solids of revolution, but it works for *all* of them.

Volume of a Solid of Revolution

If the continuous function $y = f(x)$ on the interval $a \leq x \leq b$ is rotated about the x-axis, it forms a solid. The volume of the solid is equal to $\int_a^b \pi(f(x))^2 \, dx$.

5.5.3 Surface Area of a Solid of Revolution

We can also use calculus methods to find the surface area of a solid of revolution. This requires viewing the surface as being formed by "rings" in the shape of frustums of cones. In figure 5.11, the lateral surfaces of a cone and three frustums of cones have been used to approximate the surface of a hemisphere.

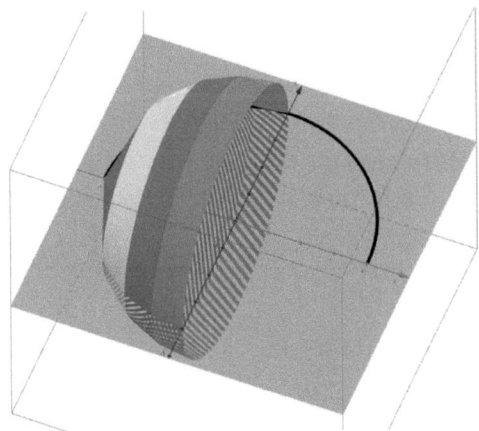

FIGURE 5.11. Approximating the surface of a hemisphere.

The area of the lateral surface of each ring is approximately equal to the product of the circumference of the circle and a small change in the length of the curve, denoted as Δl.

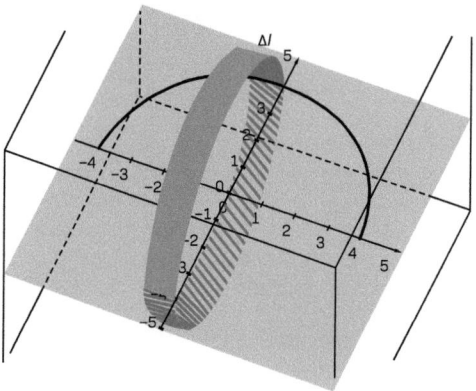

FIGURE 5.12. One ring with a width of Δl.

The circumference of the circle is $2\pi f(x)$, because $f(x)$ is the radius of the circle. As Δl gets very small, it can be approximated by $\sqrt{1+(f'(x))^2}\,\Delta x$, which we saw in section 5.4. The area of the lateral surface of one ring would then be $2\pi f(x)\sqrt{1+(f'(x))^2}\,\Delta x$. We then take the limit of the sum of these areas as the number of rings gets large and obtain the following definite integral.

> **Box 5.9: Possible Pitfalls**
>
> Here's another formula that is specifically for solids of revolution, but not other solids.

> **Surface Area of a Solid of Revolution**
>
> If the continuous, differentiable function $y = f(x)$ on the interval $a \le x \le b$ is rotated about the x-axis, it forms a solid. If $f(x)$ is positive on the interval $a \le x \le b$, then the surface area of the solid is equal to $\int_a^b 2\pi f(x)\sqrt{1+\left(f'(x)\right)^2}\,dx$

We can apply this integral to the specific case of the sphere with radius r, which may be obtained by rotating $f(x) = \sqrt{r^2 - x^2}$ on the interval $-r \le x \le r$ about the x-axis. We'll do a little preparatory work first and build our way up to $\int_{-r}^{r} 2\pi f(x)\sqrt{1+\left(f'(x)\right)^2}\,dx$

The definite integral for surface area uses the derivative, so if $f(x)=\sqrt{r^2-x^2}=(r^2-x^2)^{\frac{1}{2}}$, then the derivative[1] is $f'(x)=\frac{1}{2}(r^2-x^2)^{-\frac{1}{2}}(-2x)=\frac{-x}{\sqrt{r^2-x^2}}$.

Furthermore, $(f'(x))^2 = \left(\frac{-x}{\sqrt{r^2-x^2}}\right)^2 = \frac{x^2}{r^2-x^2}$.

We now determine $1+(f'(x))^2 = 1+\frac{x^2}{r^2-x^2} = \frac{r^2-x^2}{r^2-x^2}+\frac{x^2}{r^2-x^2} = \frac{r^2-x^2+x^2}{r^2-x^2} = \frac{r^2}{r^2-x^2}$,

which means that $\sqrt{1+(f'(x))^2} = \sqrt{\frac{r^2}{r^2-x^2}} = \frac{r}{\sqrt{r^2-x^2}}$. Now, we are ready to put this all together to find the surface area of a sphere with radius r.

$$\int_{-r}^{r} 2\pi\sqrt{r^2-x^2} \cdot \frac{r}{\sqrt{r^2-x^2}} dx = \int_{-r}^{r} 2\pi r \, dx$$
$$= 2\pi rx\Big|_{-r}^{r}$$
$$= 2\pi r(r) - 2\pi r(-r)$$
$$= 2\pi r^2 + 2\pi r^2$$
$$= 4\pi r^2$$

This is one of the cases where something very complicated becomes much simpler.

Problem Set 5.5

For problems 1–4, consider the solid obtained by rotating the function about the x-axis. Complete these four steps for each problem.

a. Name the type of solid that is formed.
b. Write a definite integral that would give the volume of this solid.
c. Use technology to approximate the volume to 2 decimal places.
d. Evaluate this integral by hand to find the exact volume.

1. $g(x)=\frac{x}{3}$, $3 \leq x \leq 9$
2. $h(x)=\sqrt{4-x^2}$, $0 \leq x \leq 2$
3. $j(x) = 2$, $1 \leq x \leq 3$
4. $p(x)=\sqrt{x}$, $0 \leq x \leq 4$ (This solid is called a paraboloid.)

For problems 5–8, consider the solid obtained by rotating the function about the x-axis. Complete the following three steps for each problem.

a. Write a definite integral that would give the surface area of this solid.
b. Use technology to approximate the surface area to 2 decimal places.
c. Evaluate this integral by hand to find the exact surface area.

5. $g(x)=\frac{x}{3}$, $3 \leq x \leq 9$
6. $h(x)=\sqrt{4-x^2}$, $0 \leq x \leq 2$

---

[1] We obtain the derivative of this function using the chain rule from section 4.3.

7. $j(x) = 2, \ 1 \leq x \leq 3$
8. $p(x) = \sqrt{x}, \ 0 \leq x \leq 4$

For problems 9–12, consider the solid obtained by rotating the function about the *x*-axis. Complete these four steps for each problem.

    a. Write a definite integral that would give the volume of this solid.
    b. Use technology to approximate volume to 2 decimal places.
    c. Write a definite integral that would give the surface area of this solid.
    d. Use technology to approximate the surface area to 2 decimal places.

9. $f(x) = e^x, \ 0 \leq x \leq 1$
10. $m(x) = \ln(x), \ 1 \leq x \leq e$
11. $n(x) = \dfrac{1}{x}, \ 1 \leq x \leq 4$
12. $q(x) = 0.5^x, \ -1 \leq x \leq 2$
13. Consider three solids: a cylinder, cone, and hemisphere, each with radius *r* and height *r*.

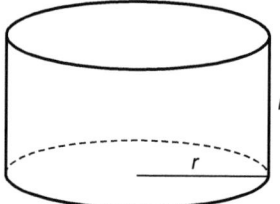 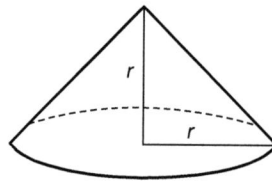 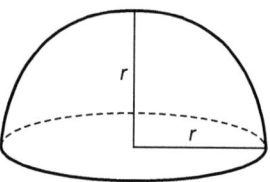

    a. Find the volume of the each of the solids in terms of *r*.

Suppose you are at the beach and you have three containers shaped like the solids.

    b. If you fill the cone with sand, how many cones of sand would you need to fill the cylinder? If you fill the cone with sand, how many cones of sand would you need to fill the hemisphere?
    c. Archie has an empty cylinder, but his cone and hemisphere are full of sand. He then pours them both into the cylinder. Which of these is the result?

- The cylinder will overflow with sand.
- The cylinder will have room for more sand.
- The cylinder will be filled exactly to the top.

Explain your answer.

References

Common Core State Standards Initiative. (n.d.). Retrieved from http://www.corestandards.org

Jones, D. (2015, April 7). Area of a circle [Video file]. Retrieved from http://youtu.be/Fl9lRHw9SDE?hd=1

Texas Education Agency. (n.d.). Texas education knowledge and skills. Retrieved from https://tea.texas.gov/curriculum/teks/

Credits

Fig. 5.5: Created with GeoGebra.
Fig. 5.11: Created with GeoGebra.
Fig. 5.12: Created with GeoGebra.

Printed by Libri Plureos GmbH in Hamburg, Germany